FOR THE
IB DIPLOMA

W9-ARM-131

Physics

Study and Revision Guide

John Allum

HODDER
EDUCATION

The Publishers would like to thank the following for permission to reproduce copyright material.

Photo credits

p.3 © Tropical Studio/Shutterstock; **p.16** © Joggie Borma/Fotolia; **p.26** © Dmitry Kalinovsky/ Shutterstock; **p.31** © WARDJet via Wikpedia Commons/ https://creativecommons.org/ licenses/by-sa/3.0/deed.en; **p.45** © Comstock Images/Getty Images; **p.52** © rvlsoft/Fotolia; **p.62** © GeoPic/Alamy Stock Photo; **p.65** © wu kailiang/Alamy Stock Photo; **p.66** © EDWARD KINSMAN/SCIENCE PHOTO LIBRARY; **p.70** © EDWARD KINSMAN/SCIENCE PHOTO LIBRARY; **p.101** *t* © Petr Bonek/Shutterstock, *b* © Image Source/Alamy Stock Photo; **p.108** © sciencephotos/Alamy Stock Photo; **p.123** © World History Archive/Alamy Stock Photo; **p.136** © Bennyartist/Fotolia; **p.138** © Rachel Wright/Alamy Stock Photo; **p.154** © University of Texas at Austin; **p.158** © Dietrich Zawischa; **p.161** © Janossy Gergely/Shutterstock; **p.165** © FooTToo/Shutterstock; **p.166** © PandaWild/Shutterstock; **p.168** © Avpics/Alamy Stock Photo; **p.198** © Gelina Semenko/Fotolia; **p.219** © BMS Education/ http://www.bmseducation.co.za; **p.224** © CHRIS PRIEST/SCIENCE PHOTO LIBRARY.

t = top, *b* = bottom, *l* = left, *r* = right, *c* = centre

Acknowledgements

The author would like to make the following acknowledgements.

Chris Talbot, formerly of Anglo-Chinese School (Independent), Singapore; Professor Robert Smith, University of Sussex (Astrophysics); Dr Tim Brown, University of Surrey (Imaging); Dr David Cooper, University of Liverpool (Quantum and Nuclear Physics).

Although every effort has been made to ensure that website addresses are correct at the time of going to press, Hodder Education cannot be held responsible for the content of any website mentioned in this book. It is sometimes possible to find a relocated web page by typing in the address of the home page for a website in the URL window of your browser.

Hachette UK's policy is to use papers that are natural, renewable and recyclable products and made from wood grown in sustainable forests. The logging and manufacturing processes are expected to conform to the environmental regulations of the country of origin.

Orders: please contact Bookpoint Ltd, 130 Milton Park, Abingdon, Oxon OX14 4SB. Telephone: (44) 01235 827720. Fax: (44) 01235 400454. Lines are open 9.00–5.00, Monday to Saturday, with a 24-hour message answering service. Visit our website at www.hoddereducation.com.

ISBN: 978 1 4718 9972 0

© John Allum 2017

First published in 2017 by Hodder Education,
An Hachette UK Company
Carmelite House
50 Victoria Embankment
London EC4Y 0DZ

www.hoddereducation.com

Impression number   5   4   3   2   1

Year   2021   2020   2019   2018   2017

Cover photo © Linden Gledhill

Illustrations by Pantek Media, Barking Dog Art, Aptara Inc.

Typeset in Goudy Old Style Std 10/12 pt by Aptara Inc.

Printed in Spain

A catalogue record for this title is available from the British Library

Contents

Introduction

This Study and Revision Guide has the following characteristics:

1. The content is in the same order as the official IB Physics syllabus wherever possible. *All* the 'Understandings' and 'Applications and skills' in the syllabus have been used among the headings. All equations and data in the IB *Physics Data booklet* have been highlighted (in bold green font) in the guide.

2. Essential knowledge has been included as *Key Concepts* (mostly) on the right hand side of the pages.

3. All other important items for revision have been included as bullet points.

4. Also included are *Common mistakes* and *Expert tips* (helpful knowledge on the 'edge' of the syllabus, that may not be essential in a first revision).

5. When scientific terms have been introduced, they have been stressed in bold orange font and are included in a *glossary*, which is in the online resources.

6. This guide contains a large number of questions within the revision material. These are designed to be a straightforward check of understanding of the concepts covered. *Detailed answers* are included in the online resources.

7. *Preparing for the IB Physics Diploma examination* has been included as an appendix in the online resources.

8. All revision material for the *Options* is to be found online.

How to use this revision guide

Preparing for an examination is not anybody's idea of fun. But it has to be done, and the process needs to be made as manageable as possible.

Everybody is different, and every student needs to develop their own best working habits. However, teachers will mostly agree on the following advice.

1. **Know the syllabus**. The headings within this study and revision guide are a close representation of the syllabus.

2. **Identify the parts of the syllabus in which you feel less confident**. Don't waste time revising things you already know well. Concentrate on your weaknesses, not your strengths.

3. **Make a schedule**. Be realistic. Do not attempt to work too long at any one time. Work at regular times.

4. **Make revision an active process**. Answering questions, writing, asking; not just reading or watching.

5. **Understanding is much more important than remembering**. If you understand a topic well, you will not need to make much effort to remember it.

6. **Know the exam**. Good exam technique is important (see online advice). Practice under examination conditions is always useful.

7. **Final revision**. This may be best as a simple review of the Key Concepts and equations highlighted in this Guide.

Measurement and uncertainties

1.1 Measurements in physics

Essential idea: Since 1948, the Système International d'Unités (SI) has been used as the preferred language of science and technology across the globe and reflects current best measurement practice.

Fundamental units

- The following are the **fundamental units** of the **SI system** of measurement:
  - ☐ mass: *kilogram* (kg)
  - ☐ length: *metre* (m)
  - ☐ time: *second* (s)
  - ☐ electric current: *ampere* (amp) (A)
  - ☐ amount of substance: *mole* (mol)
  - ☐ absolute temperature: *kelvin* (K)
  - ☐ (A seventh fundamental unit, the candela, used for light intensity, is *not* used in this course.)
- The fundamental units are all completely independent of each other, and *all* scientific measurements can be expressed in terms of these units (only). (Fundamental units are sometimes called *base units*.)
- Since all scientific measurement is based on these units, they require very precise definitions. These definitions are *not* part of the IB Physics course, except for the mole and the kelvin, which are explained in Chapter 3.

> **Expert tip**
>
> As physics has developed and measurements have become more accurate and precise, the definitions of fundamental units have evolved. The metre was originally defined as one ten-millionth of the distance from the North Pole to the equator, but that is certainly not precise enough for modern science. The metre is now defined as the length of the path travelled by light in vacuum during a time interval of $\frac{1}{299\,792\,458}$ of a second. (Neither of these definitions need to be remembered!)

Derived units

- **Derived units** are combinations of fundamental units. For example the derived unit of density is $kg\,m^{-3}$. Some derived units have their own name, for example, the unit of pressure, $N\,m^{-2}$, is called the pascal, Pa. These units are explained whenever they are introduced.
- Sometimes physicists use units that are not part of the SI system, for example years (y), electron volts (eV) or kilowatt-hours (kWh). When such non-SI units are used it is important to be able to convert between the units and the SI units for the same quantity. For example, $1\,kWh = 3.6 \times 10^6\,J$.
- The accepted format for writing derived units is, for example, $W\,m^{-2}$ for watts per square metre, and not W/m^2.

Scientific notation and metric multipliers

- Physics calculations can vary in scale from everyday life, to the incredibly small (atoms), to the astronomically large (distant galaxies).
- A consistent way of presenting data (a *scientific notation*) is needed which can cope with such enormous variations.

■ Using scientific notation and metric multipliers

- Values in science are commonly expressed using **scientific notation**, for example 3.9820×10^4, rather than $39\,820$.
- There should always be one (non-zero) digit *before* the decimal point. Zero(s) placed at the end of the number should have the same importance as any other digit.

- In everyday conversation we use words like thousand and million to help represent large numbers. Science uses a wide range of multipliers and the metric multipliers that may be used in the course are shown in Table 1.1.

Table 1.1

Prefix	Abbreviation	Value
peta	P	10^{15}
tera	T	10^{12}
giga	G	10^{9}
mega	M	10^{6}
kilo	k	10^{3}
deca	da	10^{1}
deci	d	10^{-1}
centi	c	10^{-2}
milli	m	10^{-3}
micro	μ	10^{-6}
nano	n	10^{-9}
pico	p	10^{-12}
femto	f	10^{-15}

Significant figures

Revised ☐

- Significant figures are all the digits (including zeros) used in numerical data to have meaning.
- All the digits used in scientific notation are significant figures. For example 3.48×10^6 has three significant figures.
- The number of significant figures used in data should represent the *precision* of that data.

Common mistakes

When data is not presented in scientific notation, the significance of zeros is often unclear.

■ Using SI units in the correct format for all required measurements, final answers to calculations and presentation of raw and processed data

QUESTIONS TO CHECK UNDERSTANDING

1 Express the following derived units in terms of fundamental units:
   a the newton
   b the coulomb
   c the volt
   d the radian.

2 Write the following numbers in standard notation:
   a 823.79
   b 0.0002840
   c 2

3 Convert the following into SI units:
   a 23 °C
   b 19.3 kWh
   c 38 eV
   d 50 km h^{-1}
   e 1 year.

4 a Express 2.4×10^{12} W in **i** kW, **ii** MW, **iii** GW.
   b Express in amps: **i** 347 mA, **ii** 78.4 nA.

5 Express 3.826 to:
   a 3 significant figures
   b 2 significant figures
   c 1 significant figure.

Estimation

■ Estimating quantities to an appropriate number of significant figures

■ It is an important skill to be able to make reasonable estimates of various quantities (and give them to a sensible number of significant figures), and to use such estimates to make comparisons between quantities to the nearest *order of magnitude*.

■ Orders of magnitude

■ When quantities are quoted to the nearest power of 10, it is called giving them an **order of magnitude**.

Expert tip

Consider, as examples, the numbers 279, 579 and 379. The number 279 to the nearest order of magnitude is 100 (10^2) and 579 to the nearest order of magnitude is 1000 (10^3). The number 379 is closer to 100 than 1000, so it may seem sensible to say that its nearest order of magnitude is 100 (10^2). However, since $\log 379 = 2.58$, the nearest order of magnitude is 10^3.

■ Quoting and comparing ratios, values and approximations to the nearest order of magnitude

QUESTIONS TO CHECK UNDERSTANDING

6 Without making any calculations, estimate order of magnitude values for the following:
   a the mass of a chicken egg
   b the thickness of a page in a book
   c the temperature of a flame used in cooking.

7 Estimate the volume of water in the volcanic lake shown in Figure 1.1.

Figure 1.1

8 Make calculations to estimate:
   a the pressure underneath the wheel of family car
   b the time it takes a light photon to travel across a room
   c the electrical resistance of a domestic iron.

9 Give order of magnitude ratios for the following:
   a mass of a family car / mass of a small coin
   b power provided by a large power station / power provided by a torch battery
   c the period of a long pendulum / the period of the sound from a referee's whistle.

■ Common terminology

Scientific and technological information is now quickly and easily transferred around the world, mostly using SI units and internationally agreed mathematical and scientific symbols. In earlier centuries things were very different and communication in different languages, using inconsistent symbols and units were significant problems inhibiting scientific progress.

■ Improvement in instrumentation

No matter how precise and accurate a measurement may be, there is always the possibility that future developments in instrumentation and technique will result in even greater precision. Because of this, the definitions of fundamental units have been improved in the past and this may well continue in the future.

■ Certainty

Many people believe that science deals with facts, truth and certainty. In practice, most scientists will readily admit to the opposite: they are 100% certain of very little, and *all* measurements have a measure of uncertainty.

1.2 Uncertainties and errors

Revised

Essential idea: Scientists aim towards designing experiments that can give a 'true value' from their measurements, but due to the limited precision in measuring devices, they often quote their results with some form of uncertainty.

Errors, accuracy and precision

Revised

- It is an aim of good experimental techniques and apparatus to keep errors and uncertainties as small as possible.
- We will assume that errors in measurement are due to limitations of equipment or the techniques used, and *not* due to mistakes made by the person carrying out the experiment.
- A single measurement which has only a small error is described as being **accurate**. If a set of repeated measurements of the same quantity has an average which is close to the 'true' value, then it is described as accurate, even if individual readings are not.
- In scientific research 'true' values will *not* usually be known and this means that the errors are also unknown. The accuracy of results may then be judged partly from the *uncertainty* in measurements.
- *Accuracy* and *precision* should not be confused with each other. It is possible for measurements to be precise but *not* accurate (or accurate but not precise). The difference is represented in Figure 1.2. Good experimental results are *both* accurate *and* precise.

> **Key concepts**
>
> An **error** occurs in a measurement when it is not exactly the same as the *'true'* value.
>
> A single measurement which has only a small error is described as being **accurate**.
>
> The **uncertainty** of a measurement is the range of values within which we would expect any repeated readings to occur.
>
> A measurement is described as **precise** if a similar result would be obtained if the measurement was repeated. Readings with small (random) uncertainties are precise.

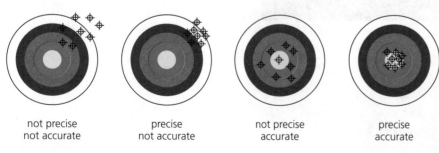

not precise	precise	not precise	precise
not accurate	not accurate	accurate	accurate

Figure 1.2

Random and systematic errors

- **Random errors** occur in *all* experiments for a variety of different reasons, but especially because of the limitations of the apparatus being used. Random errors result in measurements which are scattered (randomly) around the 'true' value.
- If the same error occurs in every measurement made using the same instrument and technique, it is called a **systematic error**. For example, this may be because a measuring instrument has a **zero offset error**. Figure 1.3 shows a zero offset error on a disconnected voltmeter.

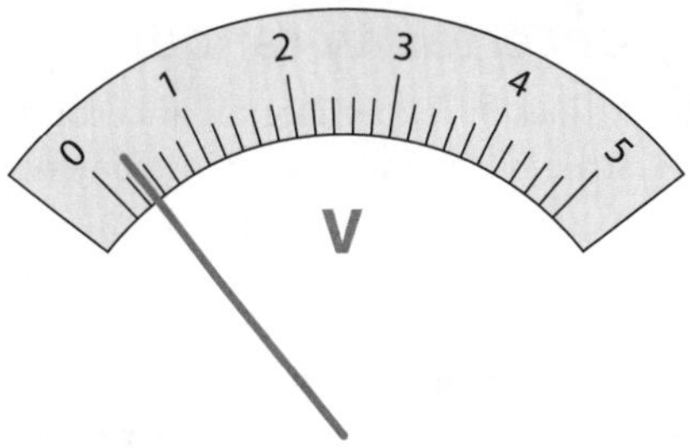

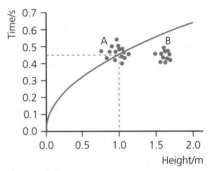

Figure 1.3

Explaining how random and systematic errors can be identified and reduced

- The curved line in Figure 1.4 shows how the *theoretical* time for an object to reach the floor varies when it is dropped from different heights.
- In an actual experiment (with heavy steel spheres), student A measured the times for a height of 1.0 m. These results show *accuracy* because the errors are small (they are close to the theoretical value), and they are *precise* (repeated results are close to each other). Because the results are scattered around the 'true' result, they are described as *random errors*.
- The results of student B's experiment for a height of 1.6 m also show random variations, and they are precise, but not accurate. *All* the readings have times which are too small compared to the theoretical value. There is a *systematic error* in all these measurements.
- The effects of *random errors* can also be reduced by taking measurements under different circumstances, so that a graph can be drawn. Drawing a *best-fit line* is a way of reducing random errors in patterns of results.
- *Systematic errors* are not reduced by simply repeating readings. Sometimes systematic errors can be identified from the pattern of results. For example, Figure 1.5 shows the results of an experiment measuring the speed of a trolley rolling down a hill from rest. The results have produced the expected straight-line graph, but unexpectedly it does not pass through the origin, so that a systematic error is probable.
- If a systematic error is a result of a zero offset error (see above), the measurements can be adjusted by adding or subtracting the error from all measurements.

Figure 1.4

> ### Key concepts
>
> The effects of *random errors* can be reduced by taking an average of repeated measurements.
>
> *Systematic errors* result in measurements which are always too high or too low. They are produced by faulty apparatus or repeating poor experimental technique. Systematic errors can often be identified from graphs of results.

QUESTIONS TO CHECK UNDERSTANDING

10 The diameter of a ball was measured five times. If the actual diameter was 19.2 cm, suggest a set of five measurements which could be described as accurate but not precise.

11 If the voltmeter shown in Figure 1.3 recorded a voltage of 5.7 V in an experiment, what was the true value of the p.d. being measured?

12 Consider the experimental results represented in Figure 1.5. Suggest a possible reason why there is a systematic error in the measurements.

13 The melting point of pure ice was measured six times with the same thermometer: 0.3 °C, 0.2 °C, 2.1 °C, 0.1 °C, 0.2 °C, 0.1 °C. Discuss the errors in these measurements.

Figure 1.5

Uncertainties

- All experimental and observational data should have their uncertainties made clear to the reader, although this is often omitted for the sake of brevity and simplicity.

◾ Collecting data that include absolute and/or fractional uncertainties and stating these as an uncertainty range (expressed as: best estimate ± uncertainty range)

- Values for experimental uncertainty depend on the smallest division of the scale of the measuring instrument and any limitations of the apparatus or experimental techniques being used. For example, the precision produced by an electronic stopwatch should be good, but when used by hand, the uncertainty in the results may be large because of the problems of starting and stopping the stopwatch at the right instants.
- Determining the uncertainty of a measurement often involves considering the pattern of results from repeated measurements, but it is common to see the uncertainty of a *single* measurement quoted, equal to the smallest division of the scale of the measuring instrument.
- The number of significant figures used in data should represent their uncertainty (precision), but not necessarily their accuracy. For example, a result of 2.792 appears to be more precise and less uncertain than a result of 2.8 (but it could be wrong).
- If a measurement is stated to be, for example, 5.83 (rather than 5.8 or 5.831) and *no* uncertainty is given, it suggests that the uncertainty may be 0.01.

◾ Absolute, fractional and percentage uncertainties

- A length measurement may be recorded as 4.3 cm ± 0.1 cm, meaning that repeated measurements would be expected to fall within the range 4.2–4.4 cm, with an average of 4.3 cm.
- Sometimes it may be appropriate to express uncertainties as *fractional uncertainties* or *percentage uncertainties*. For example if a current was measured to be 3.62 A ± 0.2 A, the fractional uncertainty is $\frac{0.2}{3.62} = 0.055$, which is equivalent to 5.5%.
- Uncertainties in trigonometric or logarithmic functions are *not* required in the IB Physics course.

> **Key concepts**
>
> Uncertainties are usually quoted as **absolute uncertainties** in the unit of measurement, with one significant digit.
>
> **Fractional uncertainties** (or **percentage uncertainties**) are usually needed when propagating uncertainties through calculations.

QUESTIONS TO CHECK UNDERSTANDING

14 The thickness of 80 sheets of paper was measured to be 0.93 cm.
  a If the smallest division on the measuring instrument was 0.1 mm, what was the percentage uncertainty in the measurement?
  b What was the thickness of one sheet and its absolute uncertainty?

15 A time measurement was recorded as 4.32 s ± 2%. Calculate:
  a the fractional uncertainty
  b the absolute uncertainty.

16 Standard laboratory 100 g masses were weighed on an electronic balance and their masses were found to be 99.5 g, 100.1 g, 99.7 g, 100.0 g and 100.2 g.
  a What was **i** the maximum absolute uncertainty in the nominal mass, **ii** the percentage uncertainty in nominal mass?
  b Determine the average mass.
  c Suggest how the manufacturers should describe their masses.

◾ Propagating uncertainties through calculations involving addition, subtraction, multiplication, division and raising to a power

- In general, a calculated result should not have more significant figures than the *least* precise data used in the calculation. For example: a power could be calculated from $P = \frac{mgh}{t} = \frac{(5.1 \times 9.81 \times 0.176)}{4.79} = 1.838299791$ W, using all the figures displayed on the calculator. However, the value 5.1, having two significant figures, is the least precise measurement used, so the calculated answer should also have only two significant figures (1.8 W).

- When making calculations based on experimental measurements (**raw data**) with known uncertainties, we need to know how to *propagate* (transfer) those uncertainties through to the final answer.
- **Addition and subtraction of similar quantities**: the *absolute* uncertainties are simply added: If $y = a \pm b$, then the uncertainty in y, $\Delta y = \Delta a + \Delta b$.
  - ☐ For example, if $a = 3.8\,cm \pm 0.1\,cm$ and $b = 12.3\,cm \pm 0.5\,cm$, then the uncertainty in $(a + b)$ *or* $(a − b)$ is $\pm(0.1 + 0.5) = \pm 0.6$. $y = a + b = 16.1 \pm 0.6\,cm$ or $y = b − a = 8.5 \pm 0.6\,cm$.
- **Multiplication or division of various quantities**: The *fractional* (or percentage) uncertainties are added to determine the fractional (or percentage) uncertainty in the result: If $y = \dfrac{ab}{c}$, then the fractional uncertainty in y, $\dfrac{\Delta y}{y} = \dfrac{\Delta a}{a} + \dfrac{\Delta b}{b} + \dfrac{\Delta c}{c}$.
  - ☐ For example, if resistance $R = \rho\dfrac{L}{A}$ with $\rho = (2.83 \pm 0.01) \times 10^{-8}\,m$, $L = 0.98 \pm 0.01\,m$ and $A = (6.78 \pm 0.05) \times 10^{-7}\,m^2$. The fractional uncertainty in R is equal to the sum of the other three uncertainties: $0.0035 + 0.0102 + 0.0074 = \pm 0.0211$ (or 2.11%).
  - ☐ Usually the *fractional* uncertainty in a calculated answer will be converted back to an *absolute* uncertainty. In the previous example the resistance can be calculated to be $0.0409056047\,2\Omega$ (using all the digits from the calculator display). Using three significant figures, the absolute uncertainty = $0.0211 \times 0.0409 = \pm 0.000\,863\,\Omega$.
  - ☐ The length (0.98 m) used in the calculation only has two significant figures and this then limits the final result, which can be expressed as $0.041\,\Omega \pm 0.001\,\Omega$. Note that the result and the uncertainty have the same number of decimal places.
- **Quantities raised to a power**: The *fractional* (or percentage) uncertainty in the result equals the fractional uncertainty in the value multiplied by the power:
  If: $y = a^n$, then $\dfrac{\Delta y}{y} = \left|\dfrac{n\Delta a}{a}\right|$ (The modulus symbol is needed because a power could be negative.)
  - ☐ For example, if $y = a^3$, and the fractional uncertainty in a is 0.06 (6%), then the fractional uncertainty in a calculated value of y is $\dfrac{\Delta y}{y} = 3 \times 0.06 = \pm 0.18$ (18%) (the same as when using the rule for $a \times a \times a$).
  - ☐ If $y = \sqrt{a}\ (= a^{\frac{1}{2}})$ and the fractional uncertainty in a is 0.12 (12%), then the fractional uncertainty in a calculated value of y is ± 0.06 (6%).

QUESTIONS TO CHECK UNDERSTANDING

17 What is the overall uncertainty in mass when masses of $2.5\,kg \pm 0.05\,kg$ and $900\,g \pm 10\,g$ are used together in an experiment?

18 The specific heat capacity of a metal can be calculated from $c = \dfrac{Q}{m\Delta T}$. Determine a value for c and its absolute uncertainty if $Q = 5.4 \times 10^3\,J \pm 9.2 \times 10^2\,J$; $m = 1.000\,kg \pm 0.005\,kg$; $\Delta T = 19\,K \pm 0.5\,K$.

19 The volume of a cube was measured to be $3.0 \pm 0.5\,cm^3$. What was the length of one side and its absolute uncertainty?

20 The time period of a mass-spring oscillating system can be calculated from $T = 2\pi\sqrt{\dfrac{m}{k}}$. Calculate a value and uncertainty for T when $m = 240\,g \pm 5\,g$ and $k = 120\,N\,m^{-1} \pm 2\,N\,m^{-1}$.

21 Explain why the percentage uncertainty of measurements often decreases with larger values.

Representing uncertainties on graphs

- Most physics investigations involve identifying two variables, and observing how they are inter-connected. The results are then plotted on a graph so that any pattern can be seen and conclusions can be reached. Figure 1.6 shows an example representing data about the motion of a train.

- Looking at this pattern of measurements, it seems obvious that there were uncertainties in the measurements, but there is no way of knowing from Figure 1.6 how large they were.

Error bars

- Absolute uncertainties in measurements are represented on graphs by *error bars* (perhaps they would be better called 'uncertainty bars'). Figure 1.7 shows the same data as Figure 1.6, but with error bars included. A *best-fit line* has been drawn which passes through all the rectangles formed by the error bars. A best-fit line like this is a way of reducing uncertainties in a pattern of results.

- The error bars for a certain quantity may all be the same, or they can sometimes vary in length. Sometimes uncertainties are too small to be represented by error bars.

> **Key concept**
> **Error bars** are vertical and horizontal lines drawn through data points to represent the magnitude of uncertainties.

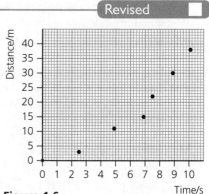

Figure 1.6

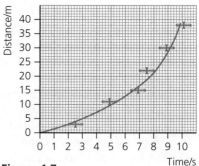

Figure 1.7

Obtaining information from graphs

- Gradients and intercepts of best-fit graphs can provide important information. Figure 1.8 shows a simple example: a best-fit straight line representing the variation of the total mass of a beaker and water as the volume of water was increased from $50\,cm^3$ to $150\,cm^3$.

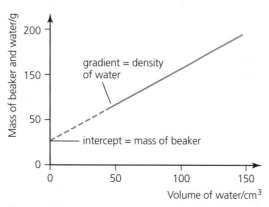

Figure 1.8

Uncertainty of gradient and intercepts

- A line midway between the lines of maximum and minimum gradient is often, but not always, the line of best fit, from which best estimates of the gradient and intercept can be determined. Figure 1.9 shows an example in which, for simplicity, the error bars are not shown.

> **Key concept**
> The uncertainty in the values of gradients and intercepts calculated from straight line graphs can be determined by comparing the lines of maximum and minimum gradient that can pass through the error bars with a line of best fit.

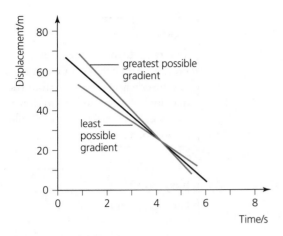

Figure 1.9

■ Determining the uncertainty in gradients and intercepts

22 What are the values of the uncertainties shown in Figure 1.7?

23 Give an example of measurements of a quantity in an experiment which have *variable* absolute uncertainties. Explain your answer.

24 Consider Figure 1.9.

  a   Estimate the gradient of the best-fit line and its uncertainty.

  b   Estimate the intercept on the time axis and its uncertainty.

NATURE OF SCIENCE

■ Uncertainties

Apart from the uncertainty in all measurements, to some extent all scientific knowledge should also be considered as uncertain. Scientists understand that there is always the possibility (perhaps small) that what is accepted knowledge today may later be found to be wrong, or a simplification of a more fundamental principle. To many, this uncertainty is a motivation and a challenge.

1.3 Vectors and scalars

Revised ☐

Essential idea: Some quantities have direction and magnitude, others have magnitude only, and this understanding is the key to correct manipulation of quantities. This sub-topic will have broad applications across multiple fields within physics and other sciences.

Vector and scalar quantities

Revised ☐

- The size of any quantity is often called its **magnitude**.
- A **vector quantity** is represented in a diagram by a straight line in the correct direction, with a length proportional to the magnitude.
- A vector −*P* has the same magnitude as the vector +*P*, but in the opposite direction.
- Examples of vectors include force, momentum and gravitational field strength.
- Examples of scalar quantities include mass, energy and time.
- If a vector, *P*, is multiplied or divided by a scalar, *k*, the result is simply *kP* or *P/k*.

> **Key concepts**
>
> A **vector** is a quantity that has both magnitude and direction.
>
> A **scalar** is a quantity that has only magnitude (no direction).

Combination and resolution of vectors

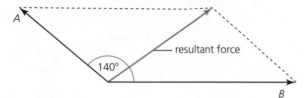

Figure 1.10

Key concept

The **resultant** of *adding* two vectors can be determined in magnitude and direction from the diagonal of a parallelogram drawn to scale, as shown by the example of two forces in Figure 1.10.

- If the angle between two vectors is 90°, the resultant can be determined algebraically.
- The *difference* between two vectors can be determined by adding the first to the negative of the second, as shown in Figure 1.11.

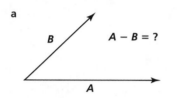
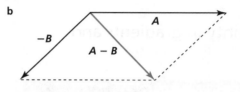

Figure 1.11

Key concept

For convenience, a single vector can be resolved into two *components* at right angles to each other. The two components, acting together, would have the same effect as the original vector.

- Sometimes a single vector does not act in a direction which is convenient for analysis. If so, the vector can be **resolved** into two **components**.
- The two components can then be considered separately and independently. Figure 1.12 is similar to a figure in the IB *Physics data booklet*. The two perpendicular components (vertical and horizontal in this example) of vector A are $A_H = A\cos\theta$ and $A_V = A\sin\theta$.

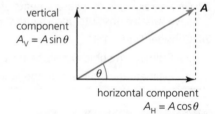

Figure 1.12

Solving vector problems graphically and algebraically

QUESTIONS TO CHECK UNDERSTANDING

25 List two vectors and two scalars (which are not mentioned above).

26 If the force to the right in Figure 1.11 is 20 N, determine the magnitude and direction of the resultant force.

27 Calculate the resultant of these two velocities: $A = 24\,\text{m s}^{-1}$ to the south and $B = 15\,\text{m s}^{-1}$ to the east.

28 What is the mathematical difference between the two vectors $(A - B)$ in the previous question?

29 A rope is used to pull a large box across a horizontal floor. The rope is pulled with a force of 247 N at an angle of 25° to the horizontal. Determine the vertical and horizontal components of this force.

NATURE OF SCIENCE

Models

Many quantities in physics need to be described by their direction (often in three dimensional space), as well as their magnitude. A branch of mathematics had to be developed to deal with these quantities and how they combine with each other: the mathematical modelling of vectors.

2 Mechanics

2.1 Motion

Revised

Essential idea: Motion may be described and analysed by the use of graphs and equations.

Distance and displacement

Revised

- **Distance**, *s*, means the length between two points. Depending on the circumstances, a quoted distance may be in a straight line or along a path of changing direction (unit: m). Distance is a scalar quantity.
- Figure 2.1 shows the path followed by some people walking around a park. The total *distance* was several kilometres, but the walk finished back at the starting point. The arrows represent the displacement from the starting point at various times. The final *displacement* was zero.

> **Key concept**
>
> **Displacement** is a vector quantity defined as the distance in a straight line from a reference point in a specified direction.

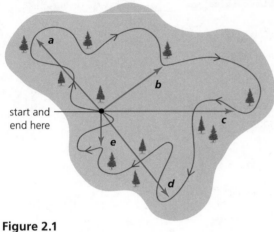

start and end here

Figure 2.1

- Displacements (and other vectors) which are in opposite directions are often represented, especially on graphs, by positive and negative signs.

Speed and velocity

Revised

- Velocity may be considered as speed in a given direction.
- Speed is a scalar quantity; velocity is a vector quantity.

> **Common mistakes**
>
> Physicists generally prefer to refer to the more precise terms of displacement and velocity, rather than distance and speed, but it is easy to confuse distance with displacement, and speed with velocity. This is sometimes because the direction involved is implied (rather than stated explicitly) and does not change. For example, it may be stated simply that a car has a *velocity* of 15 m s⁻¹, with the assumption that it is travelling in a straight line along a (given) road at a constant speed. However, if there is a reference in a question to the interaction of a moving object with other objects or forces, the vector natures of displacement, velocity, force and acceleration must be considered.

> **Key concepts**
>
> **Speed**, *v*, is defined as the rate of change of distance with time. $v = \frac{\Delta s}{\Delta t}$ (unit: m s⁻¹).
>
> **Velocity**, *v*, is defined as the rate of change of displacement with time. $v = \frac{\Delta s}{\Delta t}$ (unit: m s⁻¹).

Acceleration

- Any object that is changing the way it is moving (changing its velocity) is accelerating. This includes going faster, going slower and/or changing direction.
- An increasing velocity is known as a *positive* acceleration. A decreasing velocity is called a *negative* acceleration, or *deceleration*.

> **Key concept**
>
> **Acceleration**, a, is defined as the rate of change of velocity with time. $a = \frac{\Delta v}{\Delta t} = \frac{(v - u)}{t}$ (unit: $\mathrm{m\,s^{-2}}$).

Determining instantaneous and average values for velocity, speed and acceleration

- When we calculate an acceleration from a change of velocity/time, we are determining the *average value* acceleration during that time. However, in physics we are usually more concerned with the precise values of quantities such as speed, velocity and acceleration *at an exact instant*. These are called **instantaneous values** (rather than average values) and they can be calculated from measurements made over very short intervals of time, or from the gradients of graphs (see below).

QUESTIONS TO CHECK UNDERSTANDING

1 X has a displacement of 520 km north of Y.

   a Estimate the road distance that a car might need to travel to go from Y to X.

   b A plane takes 45 minutes to fly in a straight line from X to Y. What was its average velocity?

2 Imagine you take a bicycle ride from your home and return one hour later. Explain why it would be pointless to state a value for your average velocity.

3 The velocity of object travelling at $10\,\mathrm{m\,s^{-1}}$ west changes to $6\,\mathrm{m\,s^{-1}}$ south.

   a What is the magnitude of the change of velocity?

   b If the velocity then changes again to $6\,\mathrm{m\,s^{-1}}$ north, what is the magnitude of the second change of velocity?

4 A runner completes a 200 m race along a partly curved track in an athletics stadium in a time of 22.4 s.

   a What was the average speed?

   b Explain why the greatest instantaneous speed must have been greater than your answer to part **a**.

   c Would the magnitude of the average velocity be greater than, less than, or the same as your answer to part **a**? Explain.

Graphs describing motion

- We are mostly concerned with displacement–time, velocity–time and, less often, acceleration–time graphs.

■ Sketching and interpreting motion (*s–t*) graphs

- In Figure 2.2, between points A and B an object is moving away from a reference point with a constant velocity $\left(\frac{\Delta s}{\Delta t}\right)$. Between points B and C the object is moving back towards the reference point with a slower constant velocity. The motion between points D and E represents an object with an even lower constant velocity moving away from the reference point, but in the opposite direction.

- If a speed is changing then the displacement–time graph is curved and gradients of *tangents* to the curve at any points represent the instantaneous speeds/velocities at those times.

- In Figure 2.3 the curve A represents an object moving away from a reference point with an increasing velocity (positive acceleration), the magnitude of which at any time (for example, t_1) may be determined from the gradient of

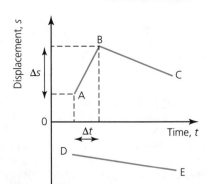

Figure 2.2

the tangent at that time $\left(\frac{\Delta s}{\Delta t}\right)$. The curve B represents an object moving in the opposite direction with a decreasing velocity (negative acceleration).

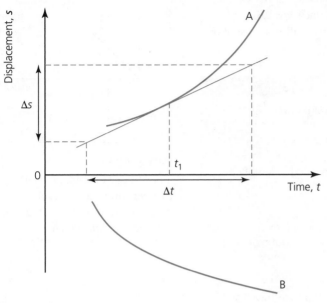

Figure 2.3

■ Sketching and interpreting motion (*v–t*) graphs

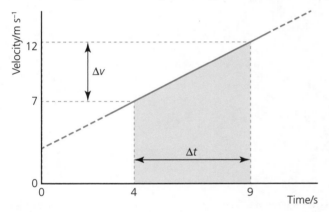

Figure 2.4

- Figure 2.4 represents a constant acceleration $\left(\frac{\Delta v}{\Delta t}\right)$. In this example $a = 1.0\,\text{ms}^{-2}$. The distance travelled between the fourth and ninth seconds can be determined from the shaded area (= 48 m to 2 sig. fig.)
- The black curved line in Figure 2.5 represents an object that is moving away from a reference point with a decreasing variable velocity. The magnitude of the acceleration at any time (like t_1) can be found from the gradient of a tangent to the curve (shown in red) at that time $\left(\frac{\Delta v}{\Delta t}\right)$. The object eventually stops moving and the total distance travelled can be determined from the shaded area under the graph.

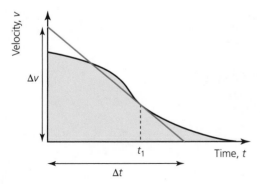

Figure 2.5

■ One common kind of motion is that of objects moving under the effects of gravity. If air resistance is negligible (sometimes called *free-fall*), all masses moving in *any* direction close to the Earth's surface accelerate downwards at the same rate. This acceleration is given the symbol *g*, and its standard value is $g = 9.81\,\text{m}\,\text{s}^{-2}$.

■ Figure 2.6 shows a velocity–time graph for an object in free-fall from rest above the Earth's surface.

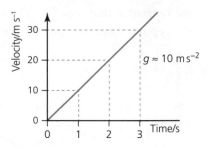

Figure 2.6

> **Key concept**
>
> The change of velocity that occurs in a certain time interval can be determined from the area under an *acceleration–time graph*.

QUESTIONS TO CHECK UNDERSTANDING

5 A car accelerates from rest at a uniform rate along a straight road. After 5.0 s its speed becomes a constant 8.0 m s⁻¹. After a further 10.0 s it begins to slow down. The rate of deceleration becomes less and less, and it finally stops 25.0 s after starting.

   a   Sketch speed–time and distance–time graphs to represent this motion.

   b   What was the car's initial acceleration?

   c   How far did the car travel in the first 15.0 s?

6 Figure 2.7 shows a displacement–time graph for a train moving along a straight track. After 4 s the train passes through a station without stopping.

   a   What was the initial velocity of the train?

   b   What was the train's velocity after 6 s?

   c   Make a copy of the graph and add a line to represent a train travelling in the opposite direction at half the speed of the first train (the trains pass at the station).

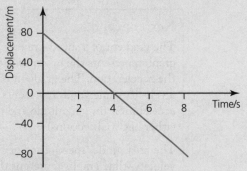

Figure 2.7

7 Figure 2.8 shows how the velocity of a falling object changed over a time of 5.0 s.

   a   Describe the motion.

   b   Determine the acceleration after 3.0 s.

   c   Estimate how far the object fell from rest in 5.0 s.

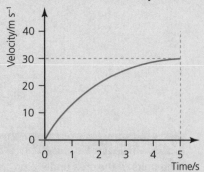

Figure 2.8

8 a   Sketch a velocity–time graph for two oscillations of a swinging pendulum.

   b   Show on your sketch an area which is equal to the amplitude of the swing.

9 Sketch a displacement–time graph for a ball dropped onto the ground, which then bounces up to two-thirds of the original height.

Equations of motion for uniform acceleration

- The simplest kind of motion to study is that of objects moving with *uniform acceleration*. The symbol u is used for the velocity at the start of observation, and the symbol v for the velocity after time t. Problems can be solved using the following **equations of motion** for uniformly accelerated motion.

- The first two equations are just definitions of acceleration and average velocity. The second two equations are mathematical combinations of the first two.
  - [ ] $v = u + at$
  - [ ] $s = \dfrac{(v + u)\,t}{2}$
  - [ ] $v^2 = u^2 + 2as$
  - [ ] $s = ut + \dfrac{1}{2}at^2$

Expert tips

It should be stressed that the equations of motion can only be used for time periods in which the accelerations remain constant. If a motion consists of two or more sections (each with its own constant acceleration), then each section must be considered separately.

Displacement, velocity and acceleration are all vectors, such that positive and negative signs may be used to represent opposite directions. For example, for an object projected vertically upwards, if the displacement is considered to be positive and increasing, while the velocity is positive and decreasing, then the acceleration (downwards) remains constant but negative.

> **Key concept**
> Given any three of the unknowns u, v, a, s, t it is possible to use the equations of motion to determine the other two.

■ Solving problems using equations of motion for uniform acceleration

QUESTIONS TO CHECK UNDERSTANDING

10 A car travelling at $13\,\mathrm{m\,s^{-1}}$ accelerates uniformly to $24\,\mathrm{m\,s^{-1}}$ in 4.6 s.
  a What was the average speed during this time?
  b How far did the car travel in the 4.6 s?
  c The car then braked with constant deceleration to stop after another 5.9 s. How far did it travel during this time?

11 A heavy sphere was dropped from rest from a height of 2.32 m.
  a What was its speed when it reached the ground?
  b What assumption did you make?

12 A ball is thrown vertically upwards from a height of 2.0 m with a speed of $18\,\mathrm{m\,s^{-1}}$.
  a Assuming that air resistance is negligible, what is its position after 3.0 s?
  b What is its velocity at the same moment?

13 Explain why the equations of motion cannot be used (on their own) to determine the speed with which a steel sphere falling through water from the surface reaches the bottom of its container.

■ Determining the acceleration of free-fall experimentally

■ If the time that a dense object takes to fall a short distance from rest can be measured accurately, then the equations of motion can be used to determine a value for the **acceleration of free-fall due to gravity**.

QUESTIONS TO CHECK UNDERSTANDING

14 a Explain why it was suggested that the object used in the acceleration of free-fall experiment should be 'dense'.

b List one advantage and one disadvantage of using greater distances for the object to fall in this experiment.

c Outline a laboratory experiment to determine the acceleration due to gravity.

15 An object falls a distance of 76.2 cm ± 0.2 cm from rest in 0.40 s ± 0.01 s. What value do these results give for '*g*' (include the absolute uncertainty)?

Fluid resistance and terminal speed

Revised ☐

■ The motion of objects through the air is opposed by the force of **air resistance**. Similar forces arise when any object moves in any direction through any fluid and generally such forces are described as **fluid resistance** or *drag*.

■ Fluid resistance arises because the fluid has to be pushed out of the path of the moving object.

■ An object which is able to move through a fluid with low resistance may be described as **streamlined**.

■ The forces acting on an object falling thorough air are shown in Figure 2.16 in Section 2.2. Similar comments can be applied to the motion of objects moving through all fluids, including liquids.

■ The skydivers in Figure 2.9a have reached their terminal speed.

Figure 2.9a

Key concept

The amount of fluid resistance acting on any moving object depends on its speed, its cross-sectional area and its shape.

When fluid resistance becomes equal and opposite to the weight, a falling object will reach a constant, **terminal speed**.

Projectile motion

Revised ☐

■ Any unpowered object moving through the air will follow a **trajectory** (path) affected by the strength of the gravitational field and (if significant) air resistance. Such objects are often called **projectiles**.

■ Analysing projectile motion, including the resolution of vertical and horizontal components of acceleration, velocity and displacement

■ We know from Section 1.3 that the instantaneous velocity of any projectile can be resolved into vertical and horizontal components ($v_v = v\sin\theta$, $v_H = v\cos\theta$). See Figure 2.9b.

Key concept

Because these components of velocity are perpendicular to each other, they can be considered independently. The fact that they do not affect each other is very useful.

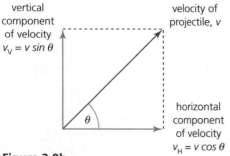

vertical component of velocity
$v_V = v\sin\theta$

velocity of projectile, *v*

horizontal component of velocity
$v_H = v\cos\theta$

θ

Figure 2.9b

■ The downwards component of velocity of an object projected horizontally will be exactly the same as for an object dropped vertically from the same height at the same time (assuming negligible air resistance). See Figure 2.10.

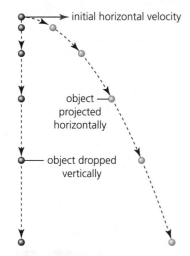

Figure 2.10

■ Projectiles move in **parabolic paths** if air resistance is negligible because the horizontal component of velocity remains constant and combines with a constant vertical acceleration.
■ The equations of motion and the conservation of energy (gravitational potential energy to or from kinetic energy) can be used with the vertical and horizontal components to predict the exact motion of a freely moving projectile. Numerical questions will assume air resistance is negligible.

■ Qualitatively describing the effect of fluid resistance on falling objects or projectiles, including reaching terminal speed

■ Air resistance is *not* usually negligible and Figure 2.11 shows its typical effects on a projectile.
■ A falling object (like the skydivers in Fig 2.9a) will reach a constant, terminal speed.

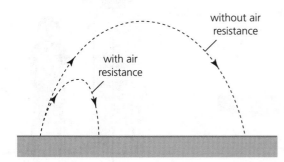

Figure 2.11

■ The effects of air resistance on objects falling vertically has already been discussed.

QUESTIONS TO CHECK UNDERSTANDING

16 A stone was thrown horizontally from a height of 1.25 m with a speed of 18 m s⁻¹. Assuming negligible air resistance, determine:
   a   the time before it reached the ground,
   b   the horizontal distance travelled before impact with the ground.

17 An air rifle at ground level fired a pellet at an angle of 20° above the horizontal. If the pellet left the rifle with a speed of 175 m s⁻¹ determine the range of the pellet (the horizontal distance to the point where it impacts the ground). Assume negligible air resistance.

18 Sketch possible displacement–time and velocity–time (0–5 s) graphs for an object dropped from rest which reaches its terminal speed after 3 s. Label the vertical scales with suggested values.

NATURE OF SCIENCE

■ Observations

Science is based on observations of the natural world and experiments. Detailed observations of various kinds of motion by scientists such as Galileo, Newton and others, started in the sixteenth and seventeenth centuries. These observations and experiments were essential for the development of many fundamental concepts in physics and science.

2.2 Forces

Revised ☐

Essential idea: Classical physics requires a force to change a state of motion, as suggested by Newton in his laws of motion.

The effects of forces

Revised ☐

■ Forces can change the motion or shapes of objects. More specifically, **resultant** *unbalanced* forces change velocities.

■ Objects as point particles

■ When considering the action of forces on various objects, in order to avoid complications, it may be convenient to assume that all of the mass of an object is effectively concentrated at one point, which is often assumed to be located at the centre of a regularly shaped object.

■ Representing forces as vectors

■ Force is a vector quantity and as such can be represented in a drawing by an arrow pointing in the right direction to, or from, the point of application. The length of the arrow should be proportional to the magnitude of the force.

■ Force is given the symbol F and it has the unit *newton*, N (explained later).

■ Figure 2.12 shows vector arrows representing the weights of two different people. The arrows have different lengths and point downwards from the centres of mass of the people.

■ Common types of *contact forces* include **reaction forces**, **tension**, **compression**, **friction**, **fluid resistance (drag)**, and **upthrust**.

> **Expert tip**
>
> When anything is pushed or pulled, the object pushes (or pulls) back, as described by Newton's third law (see below). This is called a *reaction force* and it is always perpendicular (*normal*) to the surface.

■ *Non-contact forces* are a very important part of the study of physics: **gravitational**, **electrical**, **magnetic** and **nuclear (strong and weak) forces** are all studied in this course.

■ The force of gravity pulling a mass towards a planet is called its *weight*. A vector representing weight is usually drawn acting downwards from the centre of an object (as shown in Figure 2.12).

■ Weight can be calculated from mg, because, from Newton's second law (see below), force (of weight) = mass × acceleration.

■ Also note that g may be expressed as the ratio of weight to mass. Written in this way g is known as the **gravitational field strength** and it has an accepted standard value, $g = 9.81\,\text{N kg}^{-1}$ anywhere on or close to the Earth's surface.

> **Key concepts**
>
> Resultant forces cause accelerations.
>
> It is often convenient to visualize that all of the mass of an object is at one point called its **centre of mass**.

mass 65 kg

mass 15 kg

weight, 650N weight, 150N

Figure 2.12

> **Key concept**
>
> The **weight** of an object is a gravitational force measured in newtons. It depends on the mass of the object and the strength of the gravitational field, g, in which it is located: **weight = mg**.

Free-body diagrams

- **Free-body diagrams** show all the forces acting on one object, but without showing other objects and the surroundings. It may be convenient to reduce the object to a point (as described above).
- Figure 2.13 shows a simple free-body diagram of the two forces acting on a swinging pendulum.

■ Sketching and interpreting free-body diagrams

- Examples occur throughout the course

A swinging pendulum

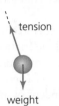

Figure 2.13

> ### QUESTIONS TO CHECK UNDERSTANDING
>
> 19 A 2.5 kg box is at rest on a slope which has an angle of 30° to the horizontal. There are three forces acting on the box. Represent these forces in a labelled free-body diagram of the box.
> 20 The gravitational field strength on Mars is 3.8 N kg⁻¹.
>   a   What would be the weight of a 620 g basketball on Mars?
>   b   What would be its acceleration if the ball was allowed to fall freely?

> **Expert tip**
>
> From Newton's third law (see below), we know that all forces occur in pairs, but each force of the pair acts on a different object. The use of free-body diagrams avoids the confusion of showing force pairs on the same drawing.

Solving problems involving forces and determining resultant force

- When more than one force acts on an object we often need to determine the overall effect. As an example, consider Figure 2.13. It would be very useful to know the value of the single force which would have the same effect as the actual forces combined. This is known as the **resultant force**. It is sometimes described as an *unbalanced force* (assuming it is not zero).
- If three forces are in equilibrium, the resultant of any two forces is equal in magnitude and opposite in direction to the third force.
- The effects of two perpendicular components can be considered independently. Taking components can be useful if a force acts on an object at an inconvenient angle, such that we may prefer to know its effects in other directions (most commonly vertical and horizontal).

> **Key concept**
>
> The resultant of two or more forces can be found using vector addition (see Section 1.3).
>
> A single force, F, can be resolved into two *components* at right angles to each other: $F \cos \theta$ and $F \sin \theta$.

> ### QUESTIONS TO CHECK UNDERSTANDING
>
> 21 a   If in Figure 2.13 the mass of the pendulum was 68 g, what was its weight?
>   b   Use a scale drawing to determine the resultant force if the tension was 0.563 N and the angle between weight and tension was 145°.
> 22 By taking components, determine the magnitude of the frictional force which is stopping the box in question 19 from slipping down the slope.

Solid friction

- **Solid friction** is a force which opposes motion between surfaces in contact. The amount of friction between two surfaces depends on the nature and roughness of the surfaces, and the *normal* force acting between them. It should be noted that, in reality, frictional forces can be unpredictable.
- Figure 2.14a shows a block and masses being pulled to the right across the surface of a table. A pair of frictional forces will occur on the surfaces of the block and the table.
- Figure 2.14b is a free-body diagram of the block, which has been simplified to a point object. The block will accelerate to the right because there is a resultant horizontal force. If more masses are added the frictional forces will increase.

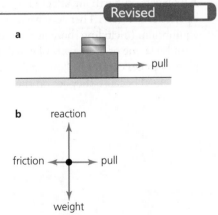

Figure 2.14

■ Describing solid friction (static and dynamic) by coefficients of friction

- We need to distinguish between the frictions that occur before and after motion begins. Before any movement occurs, we refer to **static friction**. The magnitude of a static frictional force varies up to an upper limit, just before motion begins.
- $F_f \leq \mu_s R$, where μ_s is the coefficient of static friction.
- After motion has started the friction is called **dynamic friction**. The coefficient of dynamic friction is usually less than for static friction.
- $F_f = \mu_d R$ where μ_d is the coefficient of dynamic friction. Dynamic friction may be assumed to be independent of speed.
- A common method for determining a coefficient of static friction involves placing an object on an inclined plane and increasing the angle, θ, until the object just begins to slip. At that point $\mu_s = \tan \theta$.

> **Key concept**
>
> The **coefficient of friction**, μ, is equal to the ratio:
>
> $$\frac{\text{frictional force, } F_f}{\text{normal force, } R}.$$

QUESTIONS TO CHECK UNDERSTANDING

23 Consider Figure 2.14b.

   a If the total weight was 18 N and the frictional force was 16 N, what was the coefficient of friction?

   b What kind of friction does this coefficient describe?

   c Describe what would happen if the weight was increased.

24 a A 1.2 kg wooden block was resting on an adjustable wooden sloping surface and the angle to the horizontal slowly increased. If the coefficient of static friction between the two surfaces was 0.73, calculate the angle at which the block *just* begins to slide down the slope.

   b Suggest any method by which the amount of friction between the surfaces could be reduced.

Newton's laws of motion: 1

Revised ☐

- If a resultant force acts on an object, it will accelerate.

■ Translational equilibrium

- *Translational* means moving from place to place. The motion of an object in *equilibrium* is unchanging.
- An object in *translational equilibrium* may be at rest (stationary), or moving in a straight line at a constant speed (constant velocity).

■ Describing the consequences of Newton's first law for translational equilibrium

- Newton's first law may be rephrased as: an object will remain in translational equilibrium unless a resultant force acts on it.
- All objects on Earth are affected by gravity and all moving objects are affected by frictional forces. Therefore we may assume that any object in translational equilibrium (including those at rest) *cannot* have zero forces acting on it: it must have one or more pairs of equal and opposite forces acting on it. Such pairs of forces cannot be 'force pairs in the context of Newton's third law' (see below) because they act on the same object.

> **Key concepts**
>
> **Newton's first law of motion** states that an object will remain at rest, or continue to move in a straight line at a constant speed, unless a *resultant* force acts on it.
>
> An object is in **translational equilibrium** if there is no acceleration.

> **Expert tip**
>
> It is possible that an object could be in translational equilibrium but not in rotational equilibrium, or vice versa.

■ Figure 2.15 represents a free-body diagram of a car in translational equilibrium, moving to the left with a constant velocity. There are two pairs of forces maintaining equilibrium. Note how representing the car as a point object avoids the complications of having to consider exactly how the forces are distributed (that would only need to be considered in a more detailed analysis).

> **Key concepts**
>
> Any object in *translational equilibrium* will have one or more pairs of equal and opposite forces acting on it.

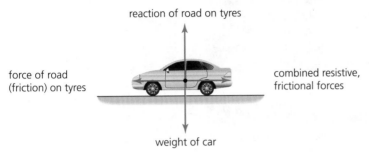

reaction of road on tyres

force of road (friction) on tyres

combined resistive, frictional forces

weight of car

Figure 2.15

■ The motion of objects falling through fluids (air in particular), has already been mentioned (Section 2.1). Figure 2.16 (b and c) shows how the two forces acting on a falling sphere vary as it accelerates from rest (Figure 2.16a).

■ The air resistance increases as the sphere moves faster, until it becomes equal and opposite to weight, as shown in Figure 2.16c. The object then falls at a constant, *terminal speed*.

■ *All* objects accelerating horizontally have a maximum speed for similar reasons: as they move faster, resistive forces increase. Eventually the resistive forces become equal to the forward force (assumed to have a limit), so that the resultant force and acceleration reduce to zero.

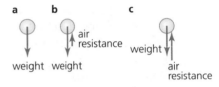

a   b   c

air resistance   weight

weight   weight   air resistance

Figure 2.16

QUESTIONS TO CHECK UNDERSTANDING

25 The Moon moves at approximately constant speed in orbit around the Earth. Is it in translational equilibrium? Explain.

26 Sketch a free-body diagram for a skydiver one second after they have jumped from a plane.

27 Describe two different ways in which a designer could increase the top speed of a car.

> **Key concept**
>
> Moving objects will reach a terminal speed when their speed has increased to a value such that fluid resistance has become equal and opposite to the force in the direction of motion.

Newton's laws of motion: 2

■ Greater masses accelerate less than smaller masses when the same (resultant) force acts on them.

■ Acceleration is proportional to resultant force (for a constant mass) and inversely proportional to mass (for a constant force).

■ One **newton** (N) is defined as that (resultant) force which accelerates 1 kg by 1 m s^{-2}.

■ (Using the concept of momentum from Section 2.4, Newton's second law can be expressed in another, more generalised, way which does not assume a constant mass: $F = \frac{\Delta p}{\Delta t}$.)

■ Using Newton's second law quantitatively and qualitatively

■ Once a resultant force has been identified, the equation $F = ma$ can be used to determine the accelerations or decelerations produced on various objects if their mass is known. Such calculations are sometimes combined with the use of the equations of motion.

> **Key concepts**
>
> **Mass** is the property of matter that resists acceleration when a (resultant) force acts.
>
> **Newton's second law** establishes a mathematical connection between (resultant) force, mass and acceleration: $F = ma$.

- More generally, in qualitative terms, it should be clear that larger forces are involved with greater accelerations or decelerations (of equal masses). During impacts (for example, consider cars colliding or people hitting the ground), any possibly harmful forces involved will be reduced if the decelerations are smaller. That is, there is less risk of injury if impacts take place over longer times and distances.

> ### Expert tip
>
> The resistance of an isolated object to any change in its motion is known as its *inertia*. The inertia of an object depends only upon its *mass*.

Common mistakes

Many people think that the safest design for a vehicle is one which is strong and very rigid. But the design of vehicles like cars involves sections of the vehicle which can collapse or crumple when large forces act on them. In this way the magnitude of the deceleration during any accident is lessened and the forces involved much reduced.

QUESTIONS TO CHECK UNDERSTANDING

28 a What average deceleration is needed in order for a plane to land on a runway in a total distance of 1.5 km if the landing speed is $75 \, \mathrm{m\,s^{-1}}$?

   b If the plane has a mass of $1.8 \times 10^5 \, \mathrm{kg}$, what average force is needed?

   c Suggest how this force is provided.

29 A bus of mass $1.68 \times 10^4 \, \mathrm{kg}$ is travelling at a constant velocity of $14.3 \, \mathrm{m\,s^{-1}}$. The forward force provided by the engine is $8.5 \times 10^3 \, \mathrm{N}$.

   a What is the magnitude of the resistive force acting on the bus?

   b What initial acceleration would be produced by increasing the forward force to $2.7 \times 10^4 \, \mathrm{N}$?

   c Assuming that the resistive force stays constant, what distance is travelled by the bus during the next 10 s?

   d Explain why, in practice, the resistive force will *not* be constant.

30 Use Newton's second law to explain why high jumpers use foam rubber to fall onto.

Newton's laws of motion: 3

Revised ☐

- This law represents the fact that *all* forces occur in pairs. For example, it is not possible to push an object unless it pushes back on you: if it cannot push back on you, then you cannot exert a force on it.
- See Figure 2.17, in which $F_A = -F_B$.

Identifying force pairs in the context of Newton's third law

- The two forces of a Newtons third law force pair *always* act on different objects. Note that the two forces are *always* of the same type as each other (e.g. both gravitational or both frictional). Clearly, Newton's third law force pairs cannot be represented on free-body diagrams because only one object is shown on such diagrams.
- Figure 2.18 shows the gravitational force pair acting on a woman and the Earth.

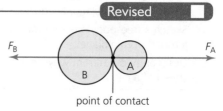

point of contact

Figure 2.17

> ### Key concept
>
> **Newton's third law** states that whenever one body exerts a force on another body, the second body exerts exactly same force on the first body, but in the opposite direction.

Expert tip

When you stand still on the floor you may be considered to be in translational equilibrium under the action of two forces: your weight downwards and the reaction force upwards from the ground on your feet. These are two different types of force acting on the same object (you), so they cannot be a Newton's third law pair. Your weight down is paired with the same sized gravitational force acting up on the Earth (which has negligible effect). These two forces are present even if there is no physical contact.

More precisely, from Chapter 6, your weight down is slightly greater than the reaction force upwards, so that there is a small net force acting on you towards the centre of the Earth. This provides the centripetal force which keeps you in circular motion on the surface of the revolving Earth.

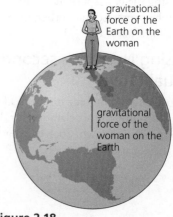

gravitational force of the Earth on the woman

gravitational force of the woman on the Earth

Figure 2.18

- Newton's third law can be very useful when explaining the propulsion of vehicles. In order to exert a forward force on the vehicle it is necessary to create a backwards force on the surroundings. For example, when the tyre on a car turns it produces a backwards force on the road surface due to friction. This is matched by a forward force from the road onto the tyre. Another example is jet propulsion: when a jet engine forces exhaust gases out backwards, an equal force is created pushing the plane forward.
- (In Section 2.4 it is explained that Newton's third law is equivalent to the law of conservation of momentum.)

QUESTIONS TO CHECK UNDERSTANDING

31 The Earth orbits around the Sun because of the force of gravity acting on it. Which other force is paired with this?

32 Figure 2.19 shows a drawing pin (thumb tack) squeezed between a person's finger and thumb.

   a Identify the four equally sized forces involved.

   b Suggest why the same force produces different sensations in the finger and thumb.

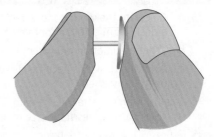

Figure 2.19

NATURE OF SCIENCE

■ Using mathematics; intuition

- The use of mathematics is an essential part of physics, but in the earlier development of the subject this was not as evident as it is now. Newton was a pioneer in this respect. His second law of motion is an obvious example, but he also deserves credit for the invention of calculus, which is essential for modern advanced physics.
- The iconic image of Newton observing an apple falling from a tree neatly represents the intuition that inspired his imagination. Newton's *Principia* is one of the most important scientific publications of all time.

2.3 Work, energy and power

Revised ☐

Essential idea: The fundamental concept of energy lays the basis upon which much of science is built.

Work done as energy transfer

Revised ☐

- **Work**, W, is the name that we give to the very common type of energy transfer that occurs when an object is displaced by a force.
- In the simplest examples, the *work done* when an object is moved can be calculated from: $W = Fs$, assuming that the force, F, is constant and *in the same direction* as the movement of the object through a distance s.
- The unit of work is the **joule**, J. 1 J is defined as the work done when a force of 1 N moves a distance of 1 m. The same unit is used with *all* forms of energy.
- Consider the equation $W = Fs \cos \theta$: If there is no movement, $s = 0$, so that no work is done by a force. If an object is moving perpendicularly to a force then $\cos \theta = 0$ and the force is not doing any work on the object. If the force and movement are in the same direction, $\cos \theta = 1$, so that the equation becomes $W = Fs$.

> **Key concept**
>
> If a constant force and the movement it produces are at an angle θ to each other (see Figure 2.20), the work done can be determined from $W = Fs \cos \theta$.

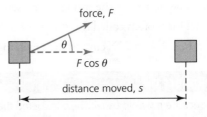

Figure 2.20

■ Determining work done including cases where a resistive force acts

- If an object is already moving, work may be done to make it move faster, gaining kinetic energy.
- Or, if there is a **resistive force** in the opposite direction to the motion (e.g. a car braking), the work done will reduce kinetic energy, usually *dissipating energy* into the surroundings as thermal energy.

QUESTIONS TO CHECK UNDERSTANDING

33 a Calculate the work done when a constant force of 150 N is used to push a desk 2.0 m across a room.

   b What assumption did you make?

   c After the desk has become stationary, to what form has the energy been transferred?

34 A car moving in a straight line with 2.0×10^5 J of kinetic energy is brought to a stop by a constant resistive force of 4.0×10^3 N. What is the stopping distance?

35 Figure 2.21 shows a case being pulled across an airport floor with a force of 18 N.

   a How much work is done in moving the case a horizontal distance of 50 m?

   b If the case has a weight of 70 N how much work is done lifting it vertically 50 cm onto a trolley?

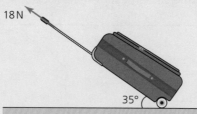

18 N

35°

Figure 2.21

Sketching and interpreting force–distance graphs

- In most practical situations forces are *not* constant, so in order to calculate work done we need to use an *average* force. If a force varies in a regular way determining an average value is not difficult $\left(\dfrac{F_{max} - F_{min}}{2}\right)$, but many forces are less predictable.

- *Force–distance graphs* can be used to show how forces vary. As an example, Figure 2.22 represents the resultant force acting on a vehicle. During section A the vehicle is being accelerated by a constant force. During section B the driver uses the brakes and there is a resultant resistive force acting on the vehicle, resulting in a deceleration. After that there is no resultant force and the vehicle will maintain its translational motion. The work done during A and/or B can be determined from the areas under the graphs.

- Another important type of force–distance graph is used to represent how materials behave when they are stretched or compressed. We refer to the resulting change of shape as a **deformation** or **strain**. Stretching a spring provides the most readily understood example. The work done is transferred to strain energy and/or internal energy. (Strain energy is discussed later in this section.)

- Figure 2.23 represents the changing length of a spring when it was stretched. The work done when the length increased from, for example, 2.2 cm to 2.6 cm can be found from the area under the graph.

- Provided that a spring is not overstretched its **extension**, Δx, (length – original length) is proportional to the force, F (Hooke's law).

- The *stiffness* of the spring is represented by the gradient of the graph. This is often called the **force constant**, k, of the spring. $k = \dfrac{\Delta F}{\Delta x}$ ($k = \dfrac{8.0}{0.4} = 20\,\text{N cm}^{-1}$ in the graph in Figure 2.23).

Key concept

The area under a force–distance graph represents the work done within the limits chosen.

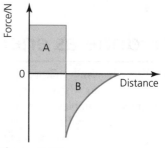

Figure 2.22

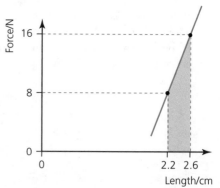

Figure 2.23

Expert tip

Because it would usually be considered that a force *causes* an extension, this graph may sometimes be drawn the other way around, with force on the horizontal axis. The gradient of a distance–force graph would then have to be interpreted differently.

QUESTIONS TO CHECK UNDERSTANDING

36 Refer to the Figure 2.23.

   a What was the original length of the spring before it was stretched (assume that it followed Hooke's law)?

   b Calculate the work done when the spring's length was increased from 2.2 cm to 2.6 cm.

37 Sketch a force–distance graph (with estimates of numerical values) to represent the impact of kicking a football.

38 Rubber becomes much stiffer the more it is stretched. This means that equal increases in extension require more and more extra force. A rubber band was stretched and it broke when its extension was 38 cm and the stretching force was 22 N.

   a Sketch a possible force–extension graph to represent this behaviour.

   b Use your graph to estimate the energy transferred to the rubber during stretching.

   c What happens to this energy if the band breaks?

> ### Key concept
> The gradient of a force-extension graph represents the force constant, k, (stiffness) of the spring or material. The area under the graph represents the energy transferred during the chosen extension.

Different forms of energy

Revised ☐

- Energy exists in different forms. Stored energy is very useful and is called **potential energy**.
- **Gravitational**, **elastic strain**, **chemical**, **electric**, **magnetic** and **nuclear** are all examples of potential energies. Energy is stored because of the forces involved in these processes.
- The other important forms of energy discussed in this course are **kinetic energy**, **electrical energy**, **thermal energy**, **radiant energy**, **(mechanical) wave energy** (including sound and water waves), **internal energy** and **rest mass energy**.
  - ☐ Mechanical waves are a combination of kinetic energy and potential energy (see Chapter 4).
  - ☐ Thermal energy flows from place to place because of a difference in temperature (see Chapter 3).
  - ☐ Electrical currents transfer energy from place to place because of a difference in potential (see Chapter 5).
  - ☐ Electromagnetic radiation, for example light, transfers radiant energy from place to place (see Chapter 4).
  - ☐ The random kinetic energy and potential energies of the molecules inside everything is called internal energy (see Chapter 3).
  - ☐ The mass of an object at rest has an equivalent amount of energy (see Chapter 7).
- Table 2.1 summarises the different forms of energy.

Table 2.1

Potential	Motion	Transfer	Molecular	Mass
gravitational		work		
electric and magnetic		mechanical waves/sound		
chemical	kinetic	electromagnetic waves	internal	rest mass
nuclear		electric currents		
elastic strain		thermal		

Common mistakes

Unfortunately, the use of the terms *heat*, *thermal energy* and *internal energy* can be confusing because various courses, books and teachers can use these terms in different and contradictory ways. In particular the word *heat* causes confusion, especially since the word is used so widely in everyday conversations. It should be clear that the energy of molecules *inside* a material (*internal energy*) is fundamentally different from energy spreading from place to place because of a temperature difference (*thermal energy*). It may be better not to call either of these *heat*. To add to the confusion a significant number of sources refer to *internal energy* as *thermal energy*.

Principle of conservation of energy

- When anything changes, energy is transformed. The study of physics involves identifying and quantifying these *energy transformations*.
- Calculating numerical values for the different forms of energy is an important part of physics.
- The principle of conservation of energy means that, in theory, we can numerically account for *all* the energy that is transferred in any process.
- The dissipation of energy results in increased internal energy and temperature in the surroundings and the spread of thermal energy.
- It should *not* be suggested that the dissipated energy has disappeared, or gone somewhere that we do not know. The use of the phrase 'lost energy' is misleading.
- In mechanical processes the dissipation of energy is due mainly to friction.

> ### Key concepts
> The **principle of conservation of energy** states that energy cannot be created or destroyed, it can only be transferred.
>
> Some *useful* energy is always **dissipated** (spread out and unrecoverable) into *useless* energy in the surroundings / environment in *all* **macroscopic** processes.

QUESTIONS TO CHECK UNDERSTANDING

39 List the energy transformations which take place in a smart phone.

40 An arrow is fired from a bow into a target. See Figure 2.24. What energy transformations are involved?

41 Beginning with the Sun, outline all the important energy transformations that have taken place in the production of oil and its subsequent use to generate electricity for our homes.

Figure 2.24

Kinetic energy

- Work has to be done to make objects move, or move faster. So that, all moving masses have energy because of their motion. This is called *kinetic energy*.
- Kinetic energy is connected to momentum, *p* (see Section 2.4), by the equation $E_K = \frac{p^2}{2m}$. This equation is most commonly used in atomic physics.

> ### Key concept
> Kinetic energy can be calculated from the mass and speed of an object: $E_K = \frac{1}{2}mv^2$

Expert tip

Note that kinetic energy depends on speed *squared*. This means that if, for example, the speed of a car increases from $5\,\mathrm{m\,s^{-1}}$ to $20\,\mathrm{m\,s^{-1}}$, its kinetic energy increases by a factor of $(\frac{20}{5})^2 = 16$. To stop a vehicle moving at twice the speed, four times as much kinetic energy has to be transferred from it. This may mean four times the braking distance.

Gravitational potential energy

- When a mass is raised away from the Earth, work has to be done *on* the mass. The force needed is considered to be equal to the weight of the object.
- Work done = $F \times s$ = weight (mg) × change of height (Δh) = $mg\Delta h$.

Expert tips

The *minimum* force needed to lift an object is equal to its weight. There would then be no resultant force on it and it could be lifted at a constant velocity. Any greater force will produce an acceleration upwards.

If an object is raised the same vertical height but at an angle less than 90° to the horizontal surface (along a slope for example), less force will be needed, but the distance moved will be greater. If there is no friction, the work done will be the same.

- **Gravitational potential energy**, E_p, is energy possessed by a mass because of its position in a gravitational field.
- When a mass is moved *away* from the Earth, the work done *on* the mass results in an *increase* in gravitational potential energy, ΔE_p. See Figure 2.25.
- When a mass moves *towards* the Earth work is done *by* the field on the mass, resulting in a *decrease* in gravitational potential energy (maybe it is transferred to kinetic energy).
- We refer to *changes* in gravitational potential energy (rather than absolute values) because a mass on a table top, or the surface of the Earth, is not considered to have zero gravitational potential energy (see Chapter 10).

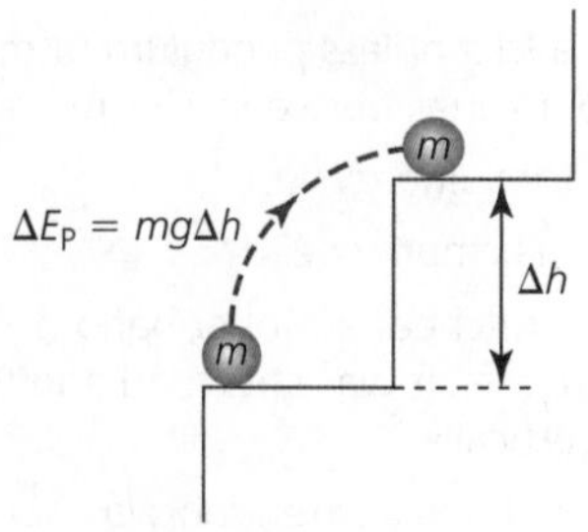

$\Delta E_P = mg\Delta h$

Δh

Figure 2.25

> **Expert tip**
>
> The equation $\Delta E_p = mg\Delta h$ can only be used for motion on, or close to, the Earth's surface, where g is constant. Chapter 10 considers calculations of gravitational potential energy in circumstance where g varies.

> **Key concept**
>
> Changes in gravitational potential energy are equal to the work done during the movement: $\Delta E_p = mg\Delta h$.

Elastic potential energy

Revised ☐

- A material can be described as **elastic** if it returns to its original shape after a deforming force has been removed. (*Elastic* does not necessarily suggest that a material is easy to stretch.)
- When a spring or material is deformed elastically we say that **elastic potential energy** is stored in it because the energy could be used to do something useful when the force is removed. A material which is deformed is said to have a *strain*, so that this kind of energy is sometimes known as *elastic strain energy*.

> **Key concepts**
>
> Elastic potential energy can be determined from $E_p = \frac{1}{2}k\Delta x^2$ where k is the force constant of the spring/material and Δx is the extension.
>
> Elastic potential energy may also be determined from the area under a force-extension graph.

Discussing the conservation of total energy within energy transformations

Revised ☐

- This heading could apply to any topic within physics, but here we are just discussing mechanical energy transformations involving forces changing the shape, position or motion of an object. The following two energy transformations are very common.
- When objects moving horizontally are accelerated (or decelerated) by a constant force, F, acting at an angle θ to the direction of motion through a distance s, the work done equals the change in kinetic energy: $Fs\cos\theta = \frac{1}{2}mv^2 - \frac{1}{2}mu^2$. As discussed before, if the force varies, an average value must be used in the calculation, which may mean that the area under a force–distance graph must be considered.
- For unpowered objects moving up or down vertically under the effects of gravity (without air resistance or friction), the change in gravitational potential energy equals the change in kinetic energy: $mg\Delta h = \frac{1}{2}mv^2 - \frac{1}{2}mu^2$.

> **Key concept**
>
> The principle of conservation of energy can be used with the equations for mechanical energies to help predict what will happen in a wide range of interactions. The results of such calculations should be considered as estimates because they do not allow for the inevitable dissipation of some energy.

QUESTIONS TO CHECK UNDERSTANDING

42 What average resultant force is needed to increase the speed of a 1.5×10^5 kg train from $10\,\text{m s}^{-1}$ to $16\,\text{m s}^{-1}$ in a distance of 2 km?

43 a Explain why a car will probably need a greater distance to decelerate from $30\,\text{m s}^{-1}$ to $20\,\text{m s}^{-1}$ than from $20\,\text{m s}^{-1}$ to $10\,\text{m s}^{-1}$.

   b After the speed of a car has been reduced, where has the difference in kinetic energy gone?

> **Expert tip**
>
> An object sliding down a slope would reach the bottom with the same speed as an object dropped vertically the same distance *if* there was no friction.

44 When a frictionless pendulum of mass 84 g swings, its vertical height above the floor varies between 1.38 m and 1.29 m. What is:

  a its total energy

  b its maximum speed?

45 When stretched, a rubber band had an effective average force constant of 420 N m⁻¹. It was stretched and then released to project a small mass of 4.3 g vertically.

  a If the band's extension was 12 cm, what is the maximum possible height to which the mass could rise?

  b Explain why, in practice, the height will be considerably less.

46 A steel sphere of mass 12 g was dropped from a height of 1.24 m onto sand.

  a What was the speed of impact?

  b The sphere was removed carefully from the sand and the depth of the indentation was measured to be 0.70 cm. Determine the average retarding force on the sphere.

  c After the impact the ball was stationary. Where has its previous energy gone?

> **Common mistakes**
>
> The last question contains an important, but widely misunderstood fact: The force that a falling object exerts on a surface is *not* equal to its weight. The magnitude of the force is dependent on the nature of the impact.

Power as rate of energy transfer

Revised

- The time taken for similar energy transfers can vary considerably.
- *Power* is the rate of doing work, or more generally, the rate of transferring energy.
- The unit of power is the **watt**, W (1 W = 1 J s⁻¹).
- Since the work done by a constant force on a vehicle moving in a straight line with a constant speed, v, is Fs, the power needed is $\frac{Fs}{t}$, so that $P = Fv$. Because the vehicle has a constant velocity, the forward force and the resistive forces are equal and opposite, so that this equation enables us to calculate the output power needed for a vehicle to maintain a constant speed against a known resistive force, F.

> **Key concept**
>
> Power, $P = \dfrac{\text{work done (energy transferred)}}{\text{time taken}}$

■ Solving problems involving power

- The concept of power may be usefully applied to almost any device, machine or process.
- We should distinguish between the total power *input* to a device and the useful power *output*.

QUESTIONS TO CHECK UNDERSTANDING

47 a How much energy is transferred by a 14 W light bulb in 2 hours?

  b Is 14 W the total power into the bulb or the useful power out?

  c What are the energy transfers involved?

48 What is the output power of an electric motor which raises an object of mass 120 kg a height of 4.3 m in 68 s?

49 What is the total resistive force opposing an aircraft travelling at 250 m s⁻¹ if the total output power of the engines is 1800 MW?

Efficiency

Revised

- Whenever energy is transferred in any macroscopic process, the energy output that is useful to us is *always* less than the total energy input.
- This is because some energy is *always dissipated* into the surroundings as increased internal energy and thermal energy spreading out. Figure 2.26 represents the energy 'flow' of a simple process.

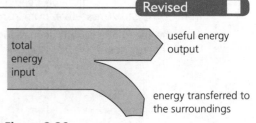

total energy input

useful energy output

energy transferred to the surroundings

Figure 2.26

- Sometimes we may refer to the 'loss' of (useful) energy, or to 'useless energy', or to energy 'wasted', but these descriptions may sometimes be considered as too vague and unhelpful.

Quantitatively describing efficiency in energy transfers

- Efficiency is a ratio, always less than one. It has no units, but it is often expressed as a percentage. That is, an efficiency of 0.50 may be quoted as 50%. Figure 2.26 illustrates a process which is about 50% efficient.
- Scientists and engineers spend a lot of time trying to improve the efficiency of processes in order to conserve energy resources and limit global warming.

> **Key concept**
>
> The **efficiency** of an energy transfer is defined as:
> efficiency =
> $$\frac{\text{useful energy (or power) output}}{\text{total energy (or power) input}}.$$

QUESTIONS TO CHECK UNDERSTANDING

50 The efficiency of a power station with a 1.8 GW output is 0.32. What is the input power?

51 a A car was accelerated from rest in order to estimate its overall efficiency. It gained kinetic energy of 4.0×10^5 J while using 45 ml of fuel. Determine the efficiency of this process if the energy density of the fuel is 3.3×10^7 J l^{-1}.

   b Would it be fair to say that the efficiency of a car travelling at constant velocity was zero?

52 Explain why an electric kettle (water heater) is much more efficient than an electric food mixer.

NATURE OF SCIENCE

Theories

The conservation of energy is one of the most important theories in the whole of science. Arising from observations and experiments over many centuries, the theory is easily expressed, but its basic generalisation is impossible to prove for all time and every possible location. However, there has *never* been any confirmed exception and scientists have even used their profound belief in the theory to explain that energy and mass are equivalent.

2.4 Momentum and impulse

FdO.

> Revised ☐

Essential idea: Conservation of momentum is an example of a law that is never violated.

Expert tip

In this chapter we are discussing only straight line (linear) motion. When objects spin they have *angular* momentum (Option B1), an important concept, but not of concern here.

Momentum

> Revised ☐

- Momentum is a vector quantity, the direction of motion must always be considered.
- Momentum is a very important concept because it enables us to interpret and use Newton's laws more broadly and, in particular, because it is always conserved in collisions (see below).

> **Key concept**
>
> The **linear momentum**, p, of a moving mass is defined as its mass × velocity (unit: kg m s^{-1}): $p = mv$.

■ As explained before, we can obtain an alternative expression for kinetic energy ($\frac{1}{2}mv^2$) using the concept of momentum: $E_K = \frac{p^2}{2m}$. This equation is used most commonly for high speed particles in atomic physics.

Newton's second law expressed in terms of rate of change of momentum

Revised ☐

■ Since $a = \frac{(v - u)}{t}$ (from Section 2.1), we can re-write Newton's second law, $F = ma$ (from Section 2.2), as $F = \frac{(mv - mu)}{t}$.

■ Using Newton's second law quantitatively and qualitatively in cases where mass is not constant

■ Newton's second law in the form $F = ma$ can only be used with constant masses moving in straight lines. If the mass is changing we need to use the law in the form $F = \frac{\Delta p}{\Delta t}$.

QUESTIONS TO CHECK UNDERSTANDING

53 a   A 250 g ball falling vertically has a speed of 12 m s^{-1} just before it hits the ground. What is its momentum?

b   What is the momentum of the ball just after it rebounds with an initial speed of 8.5 m s^{-1}?

c   What was the change in momentum of the ball?

d   If the duration of the impact was 0.37 s, what average force was exerted on the ball by the ground?

54   What is the linear momentum of an atomic particle of mass 6.7×10^{-27} kg and kinetic energy 2.7×10^{-13} J?

55   What average force is needed to accelerate a car of mass 1320 kg from 4.3 m s^{-1} to 7.5 m s^{-1} in a time of 3.9 s?

56   A space rocket is using fuel at a constant rate so that there is a constant forward force acting on it.

a   Explain why the mass of the rocket will be decreasing.

b   Use Newton's second law to account for the motion of the rocket.

Expert tip

When forces act on subatomic particles making them accelerate to very high velocities, their masses increase as well as their speeds (because of relativistic effects). This is an example of an occasion when the equation $F = ma$ would be of no use.

Key concept

Force = rate of change of momentum, $F = \frac{\Delta p}{\Delta t}$.

Conservation of linear momentum

Revised ☐

■ Newton's third law (Section 2.2) explained that if an object A exerted a force, $+F$ on object B, then object B *must* have exerted a force $-F$ on object A (the same sized force, but in the opposite direction). From the momentum interpretation of Newton's second law given above, we can now see that if, as a consequence of an isolated interaction, object A's momentum changes by $+p$, then object B's momentum *must* change by $-p$ in the same time. So that, overall, there is no change of momentum.

■ This section of the chapter concerns impacts and collisions that generally occur in short intervals of time. Generally these can be described as interactions involving forces between various objects.

■ In an *isolated* system the total momentum before any interaction is equal to the total momentum after. For masses A and B interacting (only) with each other: $p_A = -p_B$ or, in more detail: $m_A u_A + m_B u_B = m_A v_A + m_B v_B$.

■ Sometimes it may seem as if momentum (and kinetic energy) has been created from nothing. An example would be firing a gun. But it must be remembered that momentum is a vector: there was no momentum before and, when we take direction into account, the total momentum afterwards also adds up to zero.

Expert tip

It must be stressed that this analysis is only valid if there are no other forces acting. To make this clear, we often refer to *isolated systems*, or the absence of any *external forces*.

Key concept

The **law of conservation of linear momentum** states that the total (linear) momentum of a system is constant, provided that there are no external forces acting on it.

- If objects are pushed apart (quickly) we may refer to '**explosions**'. A one-dimensional example would be the recoil of a cannon when firing a cannon ball. Figure 2.27 shows the principle.

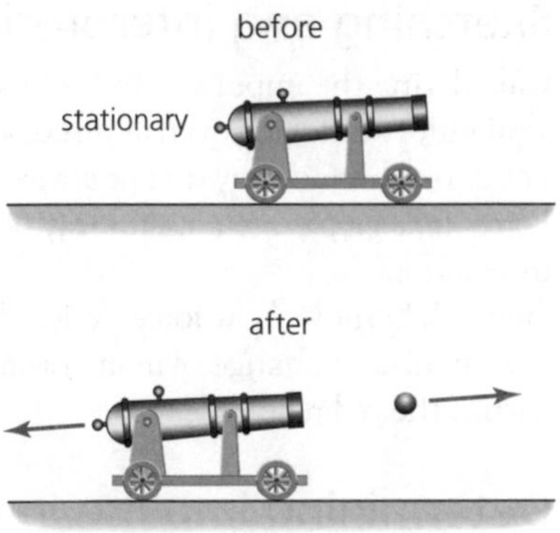

■ Applying conservation of momentum in simple isolated systems including (but not limited to) collisions, explosions, or water jets

- Since there are no exceptions, the law of conservation of momentum can be used with confidence in *any* situation to help predict what will happen in an interaction.

QUESTIONS TO CHECK UNDERSTANDING

57 A 1.2 kg trolley moving at 0.82 m s⁻¹ collides on a friction-free surface with a 1.8 kg trolley which is stationary. If the trolleys stick together, predict their speed after the collision.

58 A sphere of mass 340 g moving at 1.20 m s⁻¹ collides with another sphere of mass 220 g travelling in the opposite direction with a speed of 0.85 m s⁻¹.

   After the collision the smaller sphere returns along its original path with a speed of 0.62 m s⁻¹. What was the velocity of the larger sphere after the collision?

59 With what velocity does a 1.4 kg rifle recoil when it is firing a 3.6 g bullet at 430 m s⁻¹?

60 In a high pressure water jet cutter, such as shown in Figure 2.28, the water emerges at a speed of 700 m s⁻¹.

   a If the flow rate is 200 g min⁻¹, estimate the average force exerted while the jet impacts the metal plate.

   b If the pressure exerted by the jet is 6.0 × 10⁸ Pa, estimate the cross-sectional area of the jet.

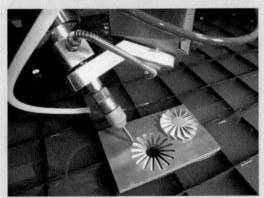

Figure 2.28

Figure 2.27

Expert tips

In macroscopic interactions a system can never be completely 'isolated' from external forces like friction. However, it is often still possible to use the conservation of momentum to help predict what will happen at the time of the interaction. However, subsequent motions would have to take into account the effect of friction, etc.

On the microscopic scale, interactions between atomic particles can usually be considered as occurring in isolated systems.

Impulse and force–time graphs

Revised

- If a force acts for a longer time it has a greater effect in changing motion. Therefore, the product of multiplying force and the time for which it acts is important, and it is called *impulse*.
- We do not use a symbol for impulse in this course. We have seen that Newton's second law can be expressed as $F = \frac{\Delta p}{\Delta t}$. Simple rearrangement gives *impulse* $= F\Delta t = \Delta p$.

Key concept

Impulse = F∆t (unit: N s).
Impulse is equal to the change of momentum.

■ Sketching and interpreting force–time graphs

- Calculating the impulse delivered by a constant force for a known time is straightforward. However, as discussed earlier, forces in interactions are usually not constant and they do not always vary in regular ways.
- *Force–time graphs* are a useful way of representing varying forces during an interaction.
- Figure 2.29 shows how force varied during a collision. The shaded area equals the impulse (= change of momentum). It can be estimated from the area within the red rectangle.

■ Determining impulse in various contexts including (but not limited to) car safety and sports

- Force–time graphs for collisions between vehicles, collisions of passengers or drivers with the interior of cars, or with seat belts (or air bags), are very useful in the analysis of safety.
- Sporting impacts, such as between a racket and a ball, can be analysed in a similar way and can lead to improved performance.

> **Key concept**
>
> The area under a force–time graph represents the impulse (= change of momentum) during an interaction.

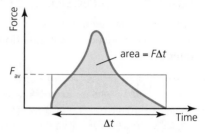

Figure 2.29

QUESTIONS TO CHECK UNDERSTANDING

61 Figure 2.30 shows how the force on a golf ball changed when it was struck by a golf club.

  a Estimate the impulse given to the ball.

  b If the ball had a mass of 46 g, determine its speed as it leaves the club.

62 a Estimate the momentum of an adult passenger in the front seat of a car moving at about 30 km h⁻¹.

  b While moving at this speed the car had an accident which quickly reduced the car's speed to zero. The passenger was protected by an air bag released during the accident. Sketch a graph to show how the force on the passenger from the air bag may have changed during this time.

  c Estimate the maximum force on the passenger. With an air bag, this force is spread over a relatively large area.

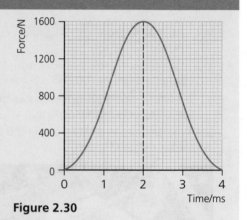

Figure 2.30

Elastic collisions, inelastic collisions and explosions

`Revised` ☐

- The use of the conservation of momentum *on its own* cannot predict the outcome of collisions. For example, when a rubber ball is dropped onto a floor the outcome is very different from the impact of an egg of the same mass and velocity with the same floor.
- To make predictions we also need to know about the nature of the colliding objects and, of course, there is a very wide range of possibilities. However, we can identify extremes: *elastic* and *inelastic* collisions.
- Perfectly elastic collisions are not possible in the macroscopic world because of the dissipation of energy into the surroundings. Collisions between particles on the atomic scale are usually (but not always) elastic.
- So, *all* realistic macroscopic collisions are inelastic: during **inelastic collisions** some of the kinetic energy is transferred to other forms of energy.
- The conservation of momentum in *explosions* has already been discussed.

> **Key concepts**
>
> A collision in which the sum of the kinetic energies of all the masses is the same after the collision as it was before is known as an **elastic collision**.
>
> When colliding objects stick together it is described as a **totally inelastic collision**.

■ Qualitatively and quantitatively comparing situations involving elastic collisions, inelastic collisions (and explosions)

Expert tips

For an *elastic collision*, as well as the conservation of momentum, we can write down an equation showing the conservation of kinetic energy. These two equations could then be solved simultaneously to determine the only possible velocities of the two masses after an elastic collision (but this is *not* required by this course). One example is well known: if a moving mass could collide elastically with an identical mass which was stationary, the moving mass would stop and the other mass would then move on with the same velocity as the first.

There are occasions when it seems as if all kinetic energy and momentum have been lost in a collision, for example when a sandbag is dropped on the floor. Examples like this involve one (or both masses) being extremely large, like the Earth itself. But momentum *must* have been conserved, which means that some kinetic energy must also still be present. However the large size of the mass means that its change of speed during the collision is unobservably small.

QUESTIONS TO CHECK UNDERSTANDING

63 A mass of 2.1 kg moving at 5.0 m s⁻¹ to the right collided on a friction-free surface with a 3.4 kg mass travelling in the opposite direction at 7.0 m s⁻¹. After the collision the smaller mass moved to the left with a speed of 9.8 m s⁻¹.

   a   Determine the velocity of the other mass.

   b   Make calculations to check if this was an elastic collision.

64 It seems that materials that behave elastically are more likely to be involved in elastic collisions. Give an example.

65 A car of mass 1340 kg travelling at 29 m s⁻¹ drove into the back of a 9600 kg truck travelling at 23 m s⁻¹ in the same direction. Assuming that there were no external forces and that this was a totally inelastic collision, determine the speed of the two vehicles immediately after the accident.

NATURE OF SCIENCE:

■ The conservation of momentum

The fact that forces must always occur in pairs (Newton's third law) can be expressed in terms of momentum conservation, a law that has profound importance because there are *no* exceptions. The law can be used to help predict events as varied as atoms emitting radiation, car accidents and asteroid impacts.

3 Thermal physics

Essential idea: Thermal physics deftly demonstrates the links between the macroscopic measurements essential to many scientific models with the microscopic properties that underlie these models.

3.1 Thermal concepts

Molecular theory of solids, liquids and gases

- **Macroscopic** describes things that we can observe with our (unaided) eyes. **Microscopic** describes things that are much smaller, and in the context of physics this often means the size of atoms and molecules.
- Most substances are made of molecules, although a few are atomic. In this chapter we use *molecules* as a general term to describe the particles in any substance.
- The molecules in *solids* are held close together by electric forces and they are usually in regular patterns. The molecules vibrate about their mean positions. See Figure 3.1.

> **Key concept**
>
> Throughout science, macroscopic observations of how substances behave can usually be explained by a microscopic understanding of what the particles within the substance are doing.

solids have a
fixed shape
and volume

liquids have a
fixed volume
but a variable shape

gases do not
have a fixed shape
or volume

molecules vibrate
in fixed position

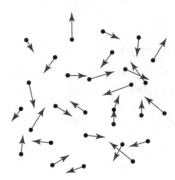

molecules have enough kinetic energy to
overcome some forces and move around

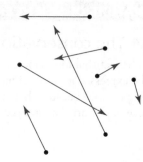

molecules move in random
directions at high speeds

Figure 3.1

- In *liquids* the molecules still vibrate, but sometimes the forces between them are overcome, allowing molecules to move around a little, but not freely. The molecules are still almost as close together as in solids, but there is little or no regularity in their arrangement, which is constantly changing.
- In *gases* the molecules are much further apart than in solids and liquids, and the forces between molecules are very, very small and considered to be negligible, except when the molecules collide. This results in all molecules moving independently in random directions with a range of different (usually fast) velocities. The velocities of molecules continually change as they collide with each other (and the walls of their container).

> **Expert tip**
>
> A typical gas density (air, for example) is about 1000 × less than the density of most liquids (water, for example). This means that each molecule has, on average, about 1000 × more space to occupy. This fact tells us that the average separation of gas molecules is about 10 × that in liquids and solids ($10^3 = 1000$). Molecular separations in solids and liquids are approximately 10^{-10} m and in gases it is about 10^{-9} m. Gas molecules are still very close together!

Internal energy

Revised ☐

- All substances contain molecules that are vibrating and/or moving around (translational motion). Because of their movement, the molecules have *kinetic energy*, which is described as *random* because there is no pattern to this movement.
- The molecules in solids and liquids also have *potential energy* (as well as kinetic energy) because there are electric forces between them. The potential energy between gas molecules is usually considered to be negligible because the molecules are much further apart.
- If the average internal energy per molecule in one part of a substance is greater than another part, then the energy will spread out until it is evenly distributed. This *movement* of energy is called *thermal energy* (see below).

> **Key concept**
> The total of all of the potential energies and the random kinetic energies of the molecules inside a substance is called its **internal energy**. (It should not be called heat)

Temperature and absolute temperature

Revised ☐

- In everyday conversation, we use the idea of **temperature** as a number which represents how hot (or cold) something is. But that is not precise enough for a scientific definition.
- We know that thermal energy flows from hotter places to colder places, so we can also say that temperatures determine the direction of net thermal energy flow.

Describing temperature change in terms of internal energy

- Since internal energy spreads out from places where it is more concentrated, we can make a connection between temperature and the concentration of energy: if the temperature of a substance increases, it is because the average internal energy of its molecules has increased.
- We will see later that temperature is related to the average random *kinetic* energy of molecules.

> **Key concept**
> When thermal energy is supplied to a substance, its internal energy increases and its temperature rises.

Using Kelvin and Celsius temperature scales and converting between them

- The **Celsius temperature scale** is an arbitrary scale based (for convenience) only on the melting and boiling points of pure water. It was *decided* to call these 0 °C and 100 °C, with one hundred divisions between them. It is important to realize that 0 °C is not in any sense a true zero of temperature.
- The *Kelvin temperature scale* was designed to overcome this problem. Its units are **kelvin**, *K*. This scale has a true zero (0 K) as the temperature at which (almost) all molecular motion has stopped (see later). For this reason the Kelvin temperature scale is also known as the **absolute temperature scale**, and 0 K is called *absolute zero*. On the Celsius scale absolute zero has the value of −273 °C (more precisely, −273.15 K).

> **Key concepts**
> (Almost) all molecular motion stops at −273° C. This is called **absolute zero**.
>
> Absolute zero is the basis for the **Kelvin temperature scale**.

- It was decided that each division on the Kelvin (absolute) scale should be equal to each division on the Celsius scale. This enables easy conversions between the two scales: **temperature (K) = temperature (°C) + 273** (see Figure 3.2). The symbol T is used to represent temperature, especially absolute temperature, but it is also common to use θ for temperatures in degrees Celsius.

- Calculations involving only *single* temperatures should always use the Kelvin scale, but calculations involving *changes* of temperature can use the Kelvin or the Celsius scale.

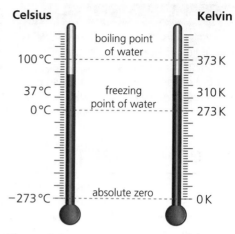

Figure 3.2

QUESTIONS TO CHECK UNDERSTANDING

1 Describe the differences in motion of a single molecule in ice, water and steam.

2 A sealed bottle of water was taken from a refrigerator and left in a warm kitchen. Explain, in terms of *its molecules*, what happened to the internal energy and temperature of the water in the next few minutes.

3 a The temperature in a room rises by 5 °C. What is this rise in kelvin?

 b The boiling point of oxygen is 90 K. Convert this temperature to °C.

4 a Explain why it may be reasonable to say that an increase in temperature from 50 K to 100 K is doubling the temperature, but increasing a temperature from 50 °C to 100 °C is *not* doubling it.

 b Why is the Kelvin temperature scale also referred to as the absolute temperature scale?

Thermal energy

Revised

- When objects are at different temperatures, energy will always flow from hotter to colder.
- The principal means of thermal energy transfer are *conduction*, *convection* and *radiation* (briefly discussed in Section 8.2).
- **Insulation** of various kinds can be used to reduce the flow of thermal energy.
- If two or more objects are at the same temperature there will be no net flow of thermal energy between them and they are described as being in **thermal equilibrium**.
- If two (or more) objects with different temperatures are able to transfer thermal energy between themselves, the cooler one will get warmer and the hotter one will get cooler, until they reach thermal equilibrium (if insulated from the surroundings), as shown in Figure 3.3.

> **Key concept**
>
> When objects are at different temperatures, energy will always flow from hotter to colder. This flow of energy is called **thermal energy** and it is given the symbol Q.

Common mistake

The same difficulties with molecular energy concepts can occur in this topic as first discussed in Topic 2. The following reminder of common mistakes is repeated from Section 2.3.

Unfortunately, the use of the terms *heat*, *thermal energy* and *internal energy* can be confusing because various courses, books and teachers can use these terms in different and sometimes contradictory ways. In particular the word *heat* causes confusion, especially since the word is used so widely in everyday conversations. It should be clear that the energy of molecules *inside* a material (*internal energy*) is fundamentally different from energy spreading from place to place because of a temperature difference (*thermal energy*). It may be better not to call either of these *heat*. To add to the confusion a significant number of sources refer to *internal energy* as *thermal energy*.

Expert tip

Thermal equilibrium should be seen as an idealized concept because perfect insulation is impossible. There will always be some flow of thermal energy to, or from, the surroundings.

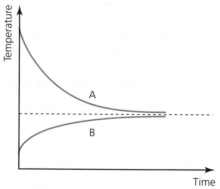

Figure 3.3

Heating

- The temperature of an object can be raised by two different methods: (1) by doing mechanical *work* on it (for example, by friction), or (2) by transferring thermal energy to it, a process we call *heating*. This chapter is mostly about the physical effects of heating.
- If a substance is heated at a *constant* rate its temperature rise every minute would be the same, *but only if* no energy was transferred to the surroundings (perfect insulation). In practice, some (possibly a lot) of the thermal energy supplied to the substance is then transferred to the surroundings.
- The dotted line in Figure 3.4 shows how the temperature of a substance heated at a constant rate changes in the idealized case when no thermal energy is transferred to the surroundings. The red line is a more realistic example: the rate of temperature rise decreases as the substance gets hotter because thermal energy is transferred away at an increasing rate. If the input power continued, eventually the temperature would become constant. Under these circumstances the substance is transferring energy to the surroundings at the same rate as energy is being supplied to it.

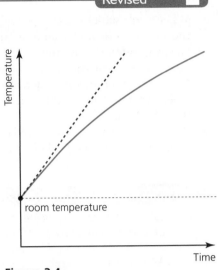

Figure 3.4

Specific heat capacity

- If we transfer the same amount of thermal energy to the same masses of different substances, the temperature rises will all be different. Each substance has a different *heat capacity*.
- If, when thermal energy Q is supplied to a mass m and the temperature rise is ΔT, the specific heat capacity of the substance can be determined from $c = \frac{Q}{m\Delta T}$. This equation is more commonly written as $Q = mc\Delta T$.
- The unit of specific heat capacity is $J\,kg^{-1}\,K^{-1}$ or $J\,kg^{-1}\,^{\circ}C^{-1}$.
- Alternatively, $mc\Delta T$ may be interpreted as the change in internal energy of a mass m of a pure substance of specific heat capacity c when its temperature changes by ΔT.

> **Key concept**
>
> The **specific heat capacity**, c, of a pure substance is defined as the amount of energy that is needed to raise the temperature of *one kilogram* by one kelvin (or 1 °C).

> **Expert tip**
>
> Raising the temperature of water is a very common human activity and it should be noted that water has a very high specific heat capacity. In practical terms this means that a relatively large amount of energy is needed to heat water, and a lot of energy has to be removed from water to cool it down.

■ Applying the calorimetric techniques of specific heat capacity experimentally

- In order to determine the specific heat capacity of a substance experimentally it is necessary to transfer a known amount of energy to a known mass, and measure the resulting temperature rise. Often, the most convenient way of doing this is by using an electrical *immersion heater* of known power.
- Figure 3.5 shows an immersion heater and thermometer inside holes drilled in an insulated metal block.
- Experiments which involve measuring thermal energy transfers may be described as **calorimetric**. (Sometimes special equipment called **calorimeters** are used.)
- In order to obtain accurate results in calorimetry experiments it is necessary to use insulation (sometimes called *lagging*) to limit thermal energy flowing to, or from, the surroundings.
- It is common to refer to the flow of energy to or from a system and its surroundings. The **system** is the object under investigation and the **surroundings** are everything else (sometimes referred to as the environment).
- Another calorimetry technique involves placing two objects or substances at different temperatures together in good *thermal* contact until they reach thermal equilibrium. The decrease in internal energy of one can then be equated to the increase in internal energy of the other, assuming that they are isolated from their surroundings.

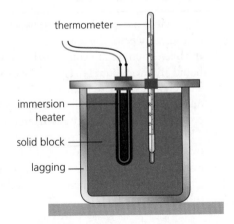

Figure 3.5

■ Calculating energy changes involving specific heat capacity

- When single substances change temperature we can use $mc\Delta T$ to calculate the energy transferred to (or from) the substance. This can be equated to the energy transferred in heating, Q, or in doing physical work, W, assuming that no energy was transferred to, or from, the surroundings.
- It is common for work to be done by resistive forces (such as friction) as the kinetic energy of an object is reduced and transferred into internal energy, so that $\Delta KE = mc\Delta T$.
- A similar calculation may be used to determine the possible temperature rise of substance (e.g. water) falling a known distance: $mg\Delta h = mc\Delta T$.

QUESTIONS TO CHECK UNDERSTANDING

5 a What is the minimum power needed for an electrical heater that is capable of raising the temperature of 0.60 kg of water from 23 °C to its boiling point in two minutes? (Assume the specific heat capacity of water is 4180 J kg^{-1} K^{-1}.)

 b Explain why, in practice, a greater power would be needed.

6 Describe an experiment to determine the specific heat capacity of water.

7 How much thermal energy has to be removed from the air in a room of volume 50 m³ in order to reduce its temperature from 30 °C to 20 °C? (Assume the specific heat capacity of air is 1000 J kg^{-1} K^{-1} and its density is 1.2 kg m^{-3}.)

8 0.56 kg of a metal alloy at 750 °C was placed in 1200 g of water at 23 °C (see Figure 3.6). The water was stirred and after thermal equilibrium had been reached the temperature of the water was 39 °C. Determine the specific heat capacity of the metal. (Assume the specific heat capacity of water is 4180 J kg^{-1} K^{-1}.)

9 Explain the difference between thermal energy and internal energy.

10 A bullet was fired at a speed of 420 m s^{-1} into a large block of wood. If 50% of the kinetic energy of the bullet was transferred to internal energy in the bullet, estimate its temperature rise. (Assume the specific heat capacity of the metal of the bullet was 480 J kg^{-1} K^{-1}.)

11 a Estimate the average temperature rise when a 10 kg bag of sand is dropped 1.6 m onto the ground. (Specific heat capacity of sand ≈ 830 J kg^{-1} K^{-1}.)

 b What assumption did you make?

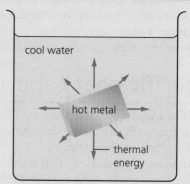

Figure 3.6

cool water

hot metal

thermal energy

Phase change

Revised

- A **phase** is a region of space in which all the physical *and* chemical properties of a substance are the same.
- In this section we are concerned with substances changing phase from solid to liquid and from liquid to gas, and vice versa.
- The change of phase from a solid to a liquid at constant temperature is called melting or **fusion**. The reverse is called **freezing** (or *solidification*). Both these processes occur at a precise temperature (for a particular substance).
- Changing from a liquid to a gas may be by the process of **boiling** or by **evaporation**. The reverse is called **condensation**.
- Evaporation can occur at any temperature (at which the substance is liquid) and it occurs only on the surface. Boiling occurs at a precise temperature throughout the liquid.

■ Describing phase change in terms of molecular behaviour

- When solids which do not melt are heated, the thermal energy supplied overcomes some intermolecular forces and therefore increases some molecular potential energies. But most of the thermal energy supplied increases the average kinetic energy of the molecules and this is observed as a rise in temperature.

Common mistakes

Do not confuse this meaning of *phase* with the use of the same word to compare oscillations (see Section 4.1).

Solids, liquids and gases are sometimes called three *states of matter*, so that changes of phase are sometimes called changes of *state*. However, it is better to avoid this term here because it may be confused with the 'state' of a gas – meaning its physical properties. To illustrate the general difference between *phase* and *state*: water and ice are two different phases and two different states (of the same substance); water and oil are also two different phases because they are different substances, but the same state.

- But at the *melting point* of the solid *all* of any thermal energy supplied is transferred to overcoming intermolecular forces and the solid's regular molecular structure is lost as the substance becomes a liquid. Since there is no increase in molecular kinetic energies, the temperature remains constant until all the solid has melted.
- Similarly, when thermal energy is supplied to a liquid at its *boiling point*, *all* of the energy is transferred to overcoming the remaining intermolecular forces as a gas is formed. Again, since there is no increase in average molecular kinetic energies, the temperature remains constant until all of the liquid has turned to gas.
- Conversely, when gases condense or liquids freeze at fixed temperatures, there are changes in molecular potential energies so that thermal energy is released, but there are no changes in molecular kinetic energies.

■ Sketching and interpreting phase change graphs

- When a hot substance (which does not change phase) is allowed to cool naturally, its temperature will fall as shown in Figure 3.7. The rate of cooling at different temperatures may be determined from gradients of the graph. The greater the difference between the temperature of the substance and the surrounding temperature, the more thermal energy per second is transferred away from the substance.
- Figure 3.8 shows how this pattern of cooling changes when there is a change of phase. This example shows a substance solidifying (above room temperature). Note that the temperature remains constant during the time taken for the solidification (as mentioned above). The same pattern would be seen in the temperature of a gas condensing to a liquid.
- Figure 3.9 shows the idealized example of a solid being heated at a constant rate (without energy losses) until it has all changed to a gas.

> **Key concept**
> The thermal energy supplied to a substance in the process of melting or boiling is used to overcome forces between molecules, but it does not increase the kinetic energy of the molecules, so that the temperature does not change.

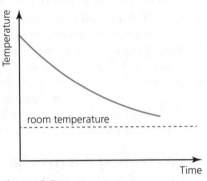

Figure 3.7

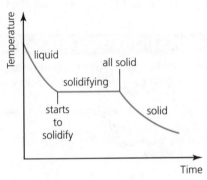

Figure 3.8

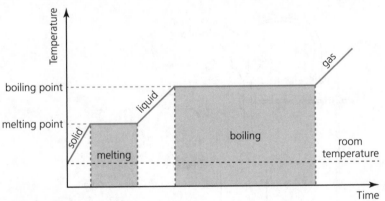

Figure 3.9

- These three change of phase graphs all have *time* on the horizontal axis, but alternatively *energy transferred* could be used instead, and the graphs would still have similar shapes.

Specific latent heat

- The energy transfer associated with a change of phase at a constant temperature is called **latent heat**. For melting or freezing this is called the **latent heat of fusion**. For boiling or condensing this is called the **latent heat of vaporization**.
- Latent heat has to be *supplied* in order to melt or boil substances, but equally important, latent heat is *emitted* from gases turning to liquids (e.g. steam condensing), or from liquids freezing.
- If energy Q changes the phase of a mass m, the latent heat can be calculated from $L = \frac{Q}{m}$ (unit: $J\,kg^{-1}$). This is more commonly written as $Q = mL$.

> **Key concept**
> The **specific latent heat** of a substance, L, is defined as the amount of thermal energy transferred when there is a change of phase of one kilogram of the substance *at constant temperature*.

Expert tip

Liquids contain molecules with a range of kinetic energies. Faster moving molecules may be able to escape from the liquid if they are close to its surface. This process is called *evaporation*. The loss of some faster moving molecules means that the *average* kinetic energy of the remaining molecules must fall. This is observed as a drop in temperature whenever evaporation occurs.

Applying the calorimetric techniques of specific latent heat experimentally

- In order to determine the latent heat of a substance experimentally, it is necessary to transfer a known amount of thermal energy to it when it is at its melting point or boiling point, and measure the corresponding mass of the substance that changed phase.
- As with other calorimetric experiments, immersion heaters can be a convenient means of providing a known thermal energy transfer, but obtaining accurate results may be difficult unless the apparatus is well insulated or the experiment is performed quickly.

Calculating energy changes involving specific latent heat of fusion and vaporization

- Freezing and boiling water are the most obvious examples.

QUESTIONS TO CHECK UNDERSTANDING

12 Figure 3.10 shows an experiment to determine the latent heat of fusion of water in the form of ice. Describe how you would carry out this experiment and use the results to determine an accurate value for the latent heat.

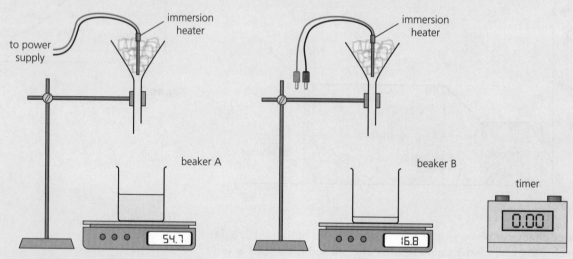

Figure 3.10

13 a How long will it take a 2.5 kW immersion heater to turn 10 g of water at 100 °C into steam at the same temperature? ($L_v = 2.26 \times 10^6$ J kg⁻¹)

   b What assumption did you make?

14 The latent heat of vaporization of a refrigerant liquid used in an air conditioner is 3.5×10^5 J kg⁻¹. What mass would have to be changed into gas (at its boiling point) to require one million joules of energy to be removed from the air?

15 a 100 g of ice at −6.0 °C was added to 400 g of water at 25.0 °C. The mixture was stirred continuously until all the ice had melted. If at that moment the temperature of the water was 4.5 °C, determine a value for the specific latent heat of fusion of water. (Assume the specific heat capacities of water and ice are 4200 J kg⁻¹ K⁻¹ and 2100 J kg⁻¹ K⁻¹, respectively.)

   b Explain why the answer to part **a** is lower than the accepted value.

■ Evidence through experimentation

The development of ideas about energy and 'heat' several hundred years ago occurred at a time when scientists did not have the instrumentation and technology that is available today. The concept that scientific advance is led by experiment and observation was not as fully established as it is now. Because of this, the particulate nature of matter was not fully understood and this lack of knowledge led to well-established but incorrect theories, which were later rejected as more experimental evidence became available.

3.2 Modelling a gas

Essential idea: The properties of ideal gases allow scientists to make predictions of the behaviour of real gases.

The physical properties of gases

- Gases are the simplest state of matter to understand. The physical behaviour of *all* gases under most circumstances follows similar patterns because the physical properties of gases depend only on the random nature of molecular motions, not chemical properties. (Refer back to the molecular theory of gases at the beginning of Section 3.1.) Forces between gas molecules are assumed to be negligible, unlike the forces between molecules in solids and liquids.
- The four macroscopic properties of a gas that we can measure are *mass*, *volume*, *temperature* and *pressure*. Together, they describe the physical state of a gas.

■ Pressure

- The equation for pressure applies to all kinds of pressure, although in this chapter we are only concerned with gas pressure.
- Gas pressure arises when a very large number of gas molecules hit the walls of a container.
- The mixture of gases in the air around us produces a large gas pressure called *atmospheric pressure* ($\approx 1.0 \times 10^5$ Pa).

> **Key concepts**
>
> **Pressure** $p = \dfrac{\text{force}}{\text{area}}$: (unit: $N\,m^{-2}$),
> 1 $N\,m^{-2}$ is called a **pascal**, Pa.
> **Density**, $\rho = \dfrac{\text{mass}}{\text{volume}}$ (unit: $kg\,m^{-3}$).

■ Investigating at least one gas law experimentally

- Assuming that experiments into the physical properties of gases are done with fixed masses of gases in sealed containers, there are just three variables which may change: *pressure*, *volume* and *temperature*.
- There are three classic experiments in which one of these three is kept constant while the relationship between the other two is investigated.
  - ☐ Keeping the volume constant, vary temperature and see what happens to the pressure (see Figure 3.11).
  - ☐ Keeping the pressure constant, vary temperature and see what happens to the volume.
  - ☐ Keeping the temperature constant, vary pressure and see what happens to the volume.
- Since all gases (or mixtures of gases) show similar patterns of physical behaviour, these experiments can be done with almost any gas or mixture of gases. Air is the obvious choice.
- The results of these classic experiments helped physicists to better understand the concepts of pressure, temperature and energy. They are known as the **gas laws**.

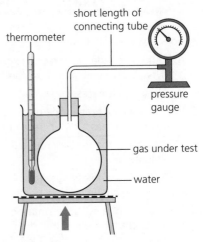

Figure 3.11

The gas laws

- The gas laws are idealized and are said to represent the behaviour of *ideal gases*. Experiments with *real gases* can produce slight variations under some circumstances (see later).
- As temperature is reduced, gas molecules move slower and each collision with the walls results in a smaller force and overall pressure.
- Results of gas laws experiments predict that *if* a gas was still a gas as it was cooled to very low temperatures, pressure and volume would reduce to (almost) zero at 0 K because all molecular motion would have stopped. For this reason 0 K is called *absolute zero*. In practice, real gases liquefy and solidify before this can happen.
- These three laws may be combined to give $pV \propto T$. This means that if a fixed mass of gas is physically changed in any way, the ratio $\frac{pV}{T}$ will always remain constant: $\frac{p_1V_1}{T_1} = \frac{p_2V_2}{T_2}$ (sometimes called the *combined gas laws equation*).

> **Key concepts**
>
> *Pressure law* – The pressure of (a fixed mass of) any gas with a constant volume is proportional to its *absolute* temperature: $p \propto T$.
>
> *Charles' law* – The volume of (a fixed mass of) any gas at constant pressure is proportional to its *absolute* temperature: $V \propto T$.
>
> *Boyle's law* – The pressure of (a fixed mass of) any gas at constant temperature is inversely proportional to its pressure: $p \propto \frac{1}{V}$.

■ Sketching and interpreting changes of state of an ideal gas on pressure–volume, pressure temperature and volume–temperature diagrams

- The three gas laws are represented graphically in the following figures (Figures 3.12, 3.13 and 3.14).

> **Expert tip**
>
> In a fourth experiment, it can be shown that the pressure of a gas is proportional to the *amount* of gas (see below), provided that the volume and temperature are unchanged.

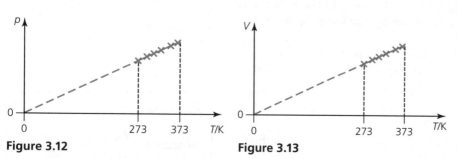

Figure 3.12   **Figure 3.13**

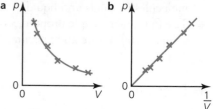

Figure 3.14

■ Solving problems using the gas laws

- For fixed masses of real gases, the equation $\frac{p_1V_1}{T_1} = \frac{p_2V_2}{T_2}$ can usually be used to predict the results of changing a gas's pressure, volume or temperature.

QUESTIONS TO CHECK UNDERSTANDING

16 Air in a container of volume 15 cm³ at a temperature of 18 °C was placed in boiling water for 10 minutes.

   a If the container was free to expand, keeping the pressure constant, what was the volume after the 10 minutes?

   b Why was it necessary to wait ten minutes?

17 In a school experiment, the pressure on a sample of air of volume 17 cm³ (and constant mass) was increased from 1.2×10^5 Pa to 2.8×10^5 Pa.

   a Assuming the temperature did not change, what was the resulting volume of the air?

   b In reality, compressing a gas will always result in a temperature rise. Suggest how it might be possible to keep the temperature rise as low as possible.

18 The volume of a cylinder containing a gas was increased from 24 cm³ to 33 cm³ and at the same time the temperature rose from 296 K to 323 K. If the original pressure was 2.3×10^5 Pa, what was the final pressure?

19 Describe in detail an experiment to investigate **i** how the volume of a gas changes when it is pressurized at constant temperature, OR **ii** how the volume of a gas changes when it is heated in a container which allows the pressure to remain constant.

20 Explain each of the three gas laws in terms of the microscopic behaviour of gas molecules.

Mole, molar mass and the Avogadro constant

Revised ☐

- The rest of Section 3.2 discusses how the unpredictable, random behaviour of individual molecules results in the predictable macroscopic properties of gases. But first we need to establish a convenient way of (indirectly) counting large numbers of molecules and relating that to masses measured in kilogrammes.
- The *amount of a substance*, n, is a measure of how many characteristic particles it contains. The amount of a substance, n, is different from its mass, m. Amount is measured in moles.
- Figure 3.15 shows three balloons of equal volumes containing three different gases at the same pressure and temperature. Each balloon contains one mole of gas (6.02×10^{23} atoms of neon or 6.02×10^{23} molecules of oxygen or carbon dioxide).

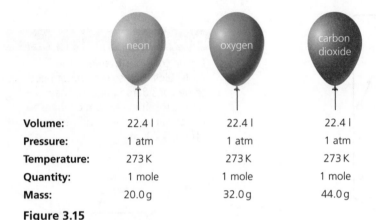

	neon	oxygen	carbon dioxide
Volume:	22.4 l	22.4 l	22.4 l
Pressure:	1 atm	1 atm	1 atm
Temperature:	273 K	273 K	273 K
Quantity:	1 mole	1 mole	1 mole
Mass:	20.0 g	32.0 g	44.0 g

Figure 3.15

- If there are N molecules in a sample of a substance, the number of moles, $n = \dfrac{N}{N_A}$.
- The molar mass of oxygen, for example, is $32.0\,\text{g mol}^{-1}$ because 32.0 g contains 6.02×10^{23} oxygen molecules, each with two atoms.
- $N = \dfrac{mN_A}{\text{molar mass}}$

> ### Key concepts
>
> The **amount of a substance**, n, is a measure of how many molecules (or atoms, if the substance is atomic) it contains.
>
> One **mole** is defined as the amount of a substance that contains the same number of molecules (or atoms) as there are atoms in exactly 12 g of carbon-12. This number is called the **Avogadro constant**, N_A, and equals $6.02 \times 10^{23}\,\text{mol}^{-1}$.
>
> **Molar mass** is defined as the mass of a substance that contains one mole (unit: g mol^{-1}).

QUESTIONS TO CHECK UNDERSTANDING

21 a How many moles are there in 1 kg of carbon-12?

   b How many atoms are in this amount?

22 A sample of hydrogen (H_2) contains 1.5×10^{24} molecules.

   a How many moles is this?

   b If the molar mass is $2.02\,\text{g mol}^{-1}$, what is the mass of this amount of hydrogen?

   c If the gas density was $2.7\,\text{kg m}^{-3}$, what was the volume of this amount of hydrogen?

23 The molar mass of carbon dioxide (CO_2) is $44.0\,\text{g mol}^{-1}$.

   a What is the mass of one molecule?

   b How many *atoms* are there in 1.0 g of the gas?

24 The molar mass of gold is $197\,\text{g mol}^{-1}$.

   a What is the mass of one atom?

   b If the density of gold is $19 \times 10^3\,\text{kg m}^{-3}$, what volume is associated with each atom?

   c Estimate the separation of gold atoms.

Kinetic model of an ideal gas

- This very important theory of classical physics uses a simplified *microscopic* model of how molecules move in a gas to mathematically explain *macroscopic* gas behaviour, including the gas laws. This simplified model predicts the behaviour of an **ideal gas**, however **real gases** usually have properties very similar to an ideal gas (see later).
- The following simplifying assumptions are made for the molecules in an ideal gas:
  - ☐ The gas is assumed to consist of a very large number of molecules that are identical and have negligible size.
  - ☐ The molecules move around randomly because there are no forces between them. Their only energy is random translational kinetic energy. See Figure 3.16.
  - ☐ Collisions between molecules and with the containing walls are elastic.

> **Key concept**
>
> The molecules in an "ideal gas" are all the same. They are considered to have negligible volume and to move around in random directions with a range of different speeds. No forces act between the molecules, except in (elastic) collisions.

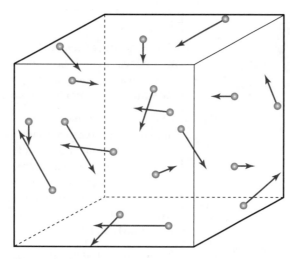

Figure 3.16

> **Expert tip**
>
> Elastic collisions between the molecules themselves do not need to be considered because they simply result in random changes to molecular velocities, with no overall effect.

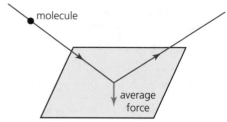

Figure 3.17

- With these assumptions we can develop a better understanding of gas pressure and temperature:
  - ☐ Each collision of a molecule with a surface results in a tiny normal force on that surface. Collisions are so frequent that the average total force on a given area (pressure) is the same throughout the gas. Figure 3.17 represents the normal force resulting from a single collision.
  - ☐ The mathematical connection between average random molecular transitional kinetic energy and temperature is $\bar{E}_K = \frac{3}{2}k_B T$. The term k_B is called the **Boltzmann constant**, it is an important constant linking macroscopic temperatures to microscopic energies. Its value is $1.38 \times 10^{-23}\,\text{J K}^{-1}$.
  - ☐ Multiplying by the Avogadro constant, N_A, gives the total random translational kinetic energy *of one mole* of any ideal gas: $\frac{3}{2}k_B N_A T$.
  - ☐ These last two equations can also be written in terms of the gas constant, R. See below.

> **Key concept**
>
> The *absolute temperature*, T, of an ideal gas is a measure of the *average* random *translational* kinetic energy of its molecules, $\bar{E}_K$.

> **Common mistake**
>
> Do not forget that the temperatures must be in kelvin. (Temperature *changes* may be in kelvin or degrees Celsius.)

Equation of state for an ideal gas

- By analysing molecular collisions with the walls in terms of Newton's laws of motion, kinetic energy of molecules and the conservation of momentum, the results of the gas law experiments can be confirmed from theory: $pV \propto T$. (This proof is not required by this course.)
- Developing the kinetic theory further leads to the equation which summarizes the physical properties of ideal gases: $pV = nRT$. This is called the **equation of state for an ideal gas**. R is known as the **(universal) gas constant**. It has the value $8.31\,\text{J K}^{-1}\,\text{mol}^{-1}$. This equation can be used for real gases under most circumstances.

> **Key concept**
>
> The equation $pV = nRT$ describes the macroscopic properties of an ideal gas. R is known as the *gas constant*.

- The macroscopic constant R is closely linked to the microscopic Boltzmann constant, k_B: $R = k_B N_A$.

- We can then express the average random translational energy of one molecule of an ideal gas in terms of R: $\bar{E}_K = \frac{3}{2}\left(\frac{R}{N_A}\right)T$.

- Remembering that the total energy of an ideal gas is only the random translational kinetic energy of its molecules, the internal energy (U) of one mole of an ideal gas equals $\frac{3}{2}RT$. And the total internal energy of any sample of an ideal gas can be determined from $\frac{3}{2}nRT$.

◼ Differences between ideal gases and real gases

- Usually real gases behave like ideal gases, except at extreme temperatures, and high pressures and densities, when the simplifying assumptions about ideal gases are no longer valid.
- An ideal gas cannot be turned into a liquid.
- Real gas molecules have rotational kinetic energy as well as translational kinetic energy.

◼ Solving problems using the equation of state for an ideal gas

- The equation $pV = nRT$ can be used with *all* real gases under most circumstances. Given any two of pressure, volume and temperature (in kelvin), the third can be calculated if the amount of gas (in moles) is also known.

Figure 3.17. Cylinders like these are used by divers. (Typically they are filled with air to a pressure about 250 x atmospheric pressure and a volume of about 12 litres.)

QUESTIONS TO CHECK UNDERSTANDING

25 a At 15°C what is the total random kinetic energy of: **i** the atoms in 1.0 mole of helium, **ii** the molecules in 1.0 mole of hydrogen?

b How much energy would need to be supplied to raise the temperature of 1.0 kg of helium in a fixed container from 15°C to 16°C? (Molar mass of helium is 4.0 g mol⁻¹.)

26 a What is the average random kinetic energy of an ideal gas molecule at 0°C?

b If the mass of oxygen molecules is 5.4×10^{-26} kg, determine their average speed at 0°C.

27 3.2 moles of an ideal gas has a volume of 120 cm³ at 58°C. What pressure is exerted on the container?

28 The molar mass of carbon dioxide is 44.0 g mol⁻¹. What mass of carbon dioxide is in a cylinder of volume 850 cm³ at 20°C if its pressure is 2.8 times greater than atmospheric pressure (1.01×10^5 Pa)?

29 Suggest why real gases behave less like ideal gases at high densities and low temperatures.

NATURE OF SCIENCE

◼ Collaboration

The kinetic model of an ideal gas was not the work of one scientist working in isolation. Sometimes, especially in the distant past, scientific advance came as the result of the insight of a great thinker (Newton for example) but, more often, especially today, the development of scientific knowledge and ideas is based on the teamwork and collaboration of many individuals, each with specialized skills and talents.

Waves

4.1 Oscillations

Essential idea: A study of oscillations underpins many areas of physics with simple harmonic motion (SHM), a fundamental oscillation that appears in various natural phenomena.

- When objects move backwards and forwards about the same place they are said to **oscillate**. Oscillations occur because there are **restoring forces** pulling or pushing displaced objects back to their **equilibrium positions**.

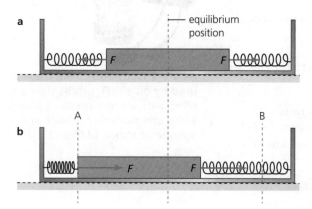

Figure 4.1

- Figure 4.1 shows a common visualization of a basic oscillation: a mass between two stretched open-wound springs on a friction-free surface. Figure 4.1a shows the mass in its equilibrium position, where there is no resultant force acting on it.
- In Figure 4.1b the mass has been displaced to position A so that there is a resultant force from the springs to the right. When released, it will accelerate, pass through the equilibrium position (where it has its maximum speed) and then decelerate to a stop at B. The motion then reverses, and so on. (In practical situations friction will eventually bring it to rest.)

Time period, frequency, amplitude, displacement and phase difference

- One **oscillation** is completed after the object next returns to the same position, moving in the same direction. One oscillation is sometimes called a **cycle**.
- Mechanical oscillations are often called **vibrations**.
- The unit of frequency is the **hertz**, Hz. 1 Hz is one oscillation per second.
- Period $= \frac{1}{\text{frequency}}$; $T = \frac{1}{f}$
- The amplitude of an oscillation is a measure of its energy. If an oscillator is given more energy its amplitude will increase.

Key concepts

For repeated oscillations each of which take the same time, one complete oscillation is completed in a time called its **period**, T.

The number of oscillations in unit time is known as the **frequency**, f.

The **displacement**, x, of an oscillator is defined as the distance from its equilibrium position in a specified direction.

Key concepts

The **amplitude** is the maximum displacement, x_0, of the oscillation.

In phase oscillators are doing exactly the same thing at the same time.

Similar oscillators with the same frequency that are not in phase are described as having a **phase difference**.

QUESTIONS TO CHECK UNDERSTANDING

1 Figure 4.2 shows two simple pendulums at the moment when they are released and allowed to swing freely.

    a Which pendulum has the greater amplitude?

    b Suggest why pendulum A has a greater time period than pendulum B.

    c What provides the restoring force which makes the pendulums oscillate?

    d Explain why it is impossible that these two pendulums can ever have a constant phase difference.

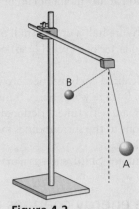

2 A mass oscillating between two springs (as in Figure 4.1) completes 50 oscillations in 84.7 s.

    a What is the time period of its motion?

    b What is its frequency?

3 Suggest a type of oscillation which does *not* keep a constant time period.

Figure 4.2

Simple harmonic oscillations

Revised ☐

- Oscillations which maintain a constant period are described as **isochronous** (or *periodic*). There is a very wide variety of such oscillations and physicists have developed a model to represent the simplest kind: *the simple harmonic oscillator*.

■ Conditions for simple harmonic motion

- If the displacement of an oscillator is doubled (for example) and, as a result, the restoring force also doubles, the acceleration back towards the equilibrium position will double ($F = ma$). This means that when the amplitude of such an oscillator is doubled it will still take the *same time* to complete an oscillation.
- More generally, the simplest kind of periodic oscillation is one which occurs when the restoring force is proportional to the displacement (but in the opposite direction). This is called simple harmonic motion (SHM).
- However, SHM is defined in terms of the oscillator rather than its cause. The negative sign in the SHM equation indicates that the acceleration is in the opposite direction to the displacement. This relationship can be represented by an acceleration–displacement graph as shown in the Figure 4.3.
- The oscillations of a mass on the end of a spring for which the extension is proportional to the force: $F = kx$ (see Section 2.2) should be simple harmonic.

Key concept

Simple harmonic motion (SHM) is defined as an oscillation in which the acceleration, a, of an object is proportional to its displacement, x, and in the opposite direction: $a \propto -x$.

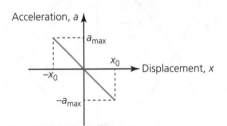

Figure 4.3

Key concept

Any graph of displacement, velocity or acceleration for SHM has the shape of a sine wave (or cosine wave). Any single graph contains the information needed to draw the other two.

■ Sketching and interpreting graphs of simple harmonic motion examples

- Graphs of displacement–time, velocity–time and acceleration–time can be drawn to represent the characteristics of SHM. Such graphs are **sinusoidal** in shape. See Figure 4.4.

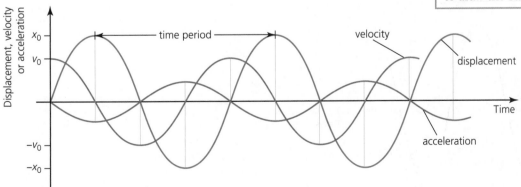

Figure 4.4

- The velocity at any moment can be found from the gradient of the displacement–time graph. The acceleration at any moment can be found from the gradient of the velocity–time graph.
- The acceleration and displacement graphs are half a cycle (π rad) out of phase with each other. The velocity graph is a quarter of a cycle ($\frac{\pi}{2}$ rad) out of phase with the other two.
- The graphs in Figure 4.4 represent the idealized circumstances in which no energy is transferred to the surroundings. If there were energy transfers out of the system, all the amplitudes would decrease with time. Frictional forces, air resistance and the transferring of vibrations to the surroundings may all result in energy being dissipated from the oscillator.
- Graphs like these can be used to represent any SHM, such as a mass oscillating on a spring, or a pendulum.

■ Qualitatively describing the energy changes taking place during one cycle of an oscillation

- Figure 4.5 shows how energies change with displacement during a simple harmonic oscillation.
- If there was no energy transferred to the surroundings, the total energy (kinetic + potential) would remain constant.
- Consider a simple pendulum as an example: at its maximum displacement the pendulum has its maximum gravitational potential energy, but no kinetic energy. As it swings towards its equilibrium position, gravitational potential energy is transferred to kinetic energy. At the centre of motion all of the pendulum's energy is kinetic. The process is then reversed, and so on.
- Figure 4.6 shows the energy changes during the first half oscillation after a pendulum was released from its highest point.

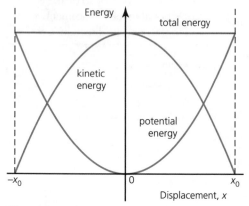

Figure 4.5

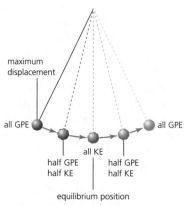

Figure 4.6

QUESTIONS TO CHECK UNDERSTANDING

4   Figure 4.7 shows how the displacement of an oscillator varied with time.
   a   Which feature of the graph suggest that this is SHM?
   b   What are the time period and frequency of the oscillation?
   c   Determine the maximum velocity of the oscillator.

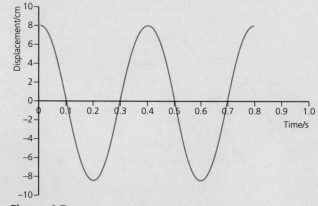

Figure 4.7

5 a Sketch a velocity–time graph for two oscillations of a pendulum undergoing simple harmonic motion. Start the graph at a time when the pendulum is at its maximum displacement.

  b Mark a point with the letter P where the pendulum is passing through its equilibrium position.

  c Another pendulum of the same frequency is swinging with a phase difference of a quarter of a cycle with the first. Represent this motion on the same axes.

6 a Sketch an acceleration–time graph for a perfect simple harmonic oscillator from a moment that it is at its maximum displacement.

  b Add a second sketch on the same axes showing how the motion would be affected if there was significant energy dissipation from the pendulum.

7 Describe the energy changes that occur during one oscillation of the simple harmonic oscillator shown in Figure 4.1.

NATURE OF SCIENCE

■ Models

SHM is a simple model in a complicated world. We are surrounded by objects which oscillate, but none of them are perfect simple harmonic oscillators. The SHM model is a good representation of a few basic oscillators (like a pendulum), but it is also the essential starting place for the analysis of more complicated oscillating systems.

4.2 Travelling waves

Revised ▢

Essential idea: There are many forms of waves available to be studied. A common characteristic of all travelling waves is that they carry energy, but generally the medium through which they travel will not be permanently disturbed.

Travelling waves

Revised ▢

- Oscillations in one part of a substance can transfer energy to their surroundings and, in this way, a continuous **travelling wave**, or a **pulse** (a wave of short duration), can travel away from its source and through the substance. Travelling waves are sometimes called *progressive* waves.
- The substance through which the wave travels is called a **medium**. The movement of a wave away from a source may be described as **propagation**. Waves involving the oscillations of masses are described as **mechanical waves**. Electromagnetic waves (see below) do not have mass and do not need a medium.
- Travelling waves transfer energy, but it is very important to note that there is no *net* motion of the medium itself in the direction of energy transfer (it just oscillates).
- Waves (in one dimension) are usually first demonstrated in school by using long ropes and/or springs.
- All travelling waves are one of only two kinds, depending upon the relative directions of their oscillations: they are either *transverse* or *longitudinal*.

> **Key concept**
> Travelling waves transfer energy away from their source.

Transverse and longitudinal waves

Revised ▢

- Electromagnetic waves and waves on stretched strings are examples of transverse waves.
- Sound is the most common example of a longitudinal wave.

■ Explaining the motion of particles of a medium when a wave passes through it for both transverse and longitudinal cases

■ Figure 4.8 shows the velocities of particles in a medium through which a transverse wave is passing from left to right. The red line represents the position of the wave a short time after the blue line.

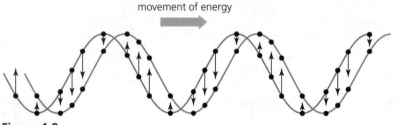

movement of energy

Figure 4.8

■ The tops and bottoms of transverse waves are often called **crests** and **troughs**.

■ The motion of particles within a medium propagating a longitudinal wave is less easy to draw, but Figure 4.9 represents a wave being sent along a slinky spring. Note that the coils of the spring are oscillating parallel to the direction in which the energy is being propagated.

■ Places where a medium is squashed (higher pressure) are called **compressions**. Places where a medium is stretched (lower pressure) are called **rarefactions**.

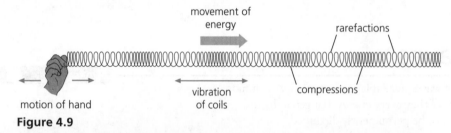

movement of energy

rarefactions

compressions

motion of hand

vibration of coils

Figure 4.9

QUESTIONS TO CHECK UNDERSTANDING

8 A table tennis ball is floating on the surface of the water in a swimming pool. At the other end of the pool a girl dives into the water, sending waves in all directions.

  a Are the waves transverse or longitudinal?

  b Describe the motion of the ball as the waves reach it.

9 When we speak we cause tiny changes to the pressure in the air. Describe, in terms of molecular movements, how sound travels from our throats to somebody's ear.

Wavelength, frequency, period and wave speed

Revised

■ Wavelength is easily shown on a displacement–position graph (see Figure 4.10).

■ Period, T, is easily represented on a displacement–time graph (see later). $T = \frac{1}{f}$.

■ Solving problems involving wave speed, frequency and wavelength

■ Since a wave travels one wavelength in one period, $c = \frac{\text{distance}}{\text{time}} = \frac{\lambda}{T}$, which is usually written as $c = f\lambda$.

■ The equation $c = f\lambda$ can be used with all kinds of waves.

QUESTIONS TO CHECK UNDERSTANDING

10 A student was observing water waves: six wave crests passed a certain point in 5.4 s. The distance from the first wave crest to the sixth was 4.80 m.

   a What was the wavelength of the waves?

   b Calculate the wave's speed.

   c Determine the frequency of the wave.

11 The frequency of the musical note middle C is 262 Hz. If the speed of sound in air is 335 m s⁻¹, what is the wavelength of this note?

12 A long stretched spring was shaken from side to side at a rate of 3.0 oscillations every second. It was estimated that the wavelength of the waves that travelled along the spring was about 40 cm.

   a What kind of wave was sent along the spring?

   b Estimate the speed of the wave.

   c As the wave travelled it transferred some energy to the surroundings. How did this affect the wave?

   d If the same spring was stretched more, it was found that the wave speed increased. Suggest a reason for this.

13 Light travels as a wave, but it has a very high frequency, typically 5×10^{14} Hz. Determine an order of magnitude for the wavelength of light waves if they travel at a speed of 3×10^8 m s⁻¹.

> ### Key concepts
> The **frequency**, f, of a wave is defined as the number of waves that pass a given point in unit time.
>
> **Wave speed**, c, is defined as the distance travelled by a wave in unit time.

> ### Expert tip
> *Amplitude* is another important measurable property of a wave. As with oscillations, amplitude means maximum displacement, and a greater amplitude means more energy. We will see later that amplitude is related to the *intensity* of a wave.

■ Sketching and interpreting displacement–distance graphs and displacement–time graphs for transverse and longitudinal waves

■ A *displacement–distance graph* for a wave shows the displacement of particles of the medium from their mean positions, and how that varies with distance from a reference point (maybe the source of the waves). If the wave is transverse, the shape of the graph may be considered to be like a 'snapshot'. See Figure 4.10.

■ All points on the wave are oscillating in the same way, but there are phase differences. As already stated, the shortest distance between two points moving in phase is one wavelength.

> ### Key concept
> Graphs used to represent the displacement of a medium as a wave passes through it are sinusoidal in shape. It is easy to get displacement–time graphs and displacement–distance graphs confused because they look alike.

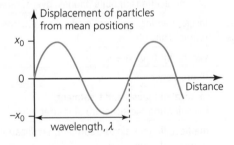

Figure 4.10

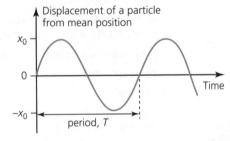

Figure 4.11

■ It is important to understand that this kind of graph can be used to represent transverse *and* longitudinal waves.

■ Wavelength and amplitude are easily shown on this kind of graph.

■ A *displacement–time graph* for a wave shows how the displacement of a particular particle/location within the medium changes with time. Again, it may represent either kind of wave. See Figure 4.11.

■ Period and amplitude are easily shown on this kind of graph.

> ### Common mistake
> Figure 4.10 might appear to only represent a transverse wave because the displacement axis is perpendicular to the distance axis. However, the graph does not show the *direction* of the displacement, which could also be parallel to the wave direction, which would then represent a longitudinal wave.

QUESTIONS TO CHECK UNDERSTANDING

14 Make a copy of Figure 4.10 and add a sketch to show the displacement half a period later.

15 Sketch a displacement–time graph to show 3 oscillations for a longitudinal wave of amplitude 2 cm and frequency of 5 Hz.

The nature of electromagnetic waves

■ The waves of the **electromagnetic spectrum** are very important in physics. Of course, we are most familiar with a small part of the overall spectrum, visible light, which has a continuous range of colours from red to violet. See Figure 4.44 in Section 4.4 for how a triangular **prism** creates a spectrum from white light.

> **Key concept**
>
> All **electromagnetic waves** consist of oscillating electric and magnetic fields which are perpendicular to each other and do not need a medium through which to travel. See Figure 4.12

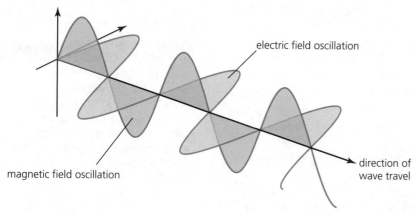

electric field oscillation

magnetic field oscillation

direction of wave travel

Figure 4.12

■ All electromagnetic waves travel across **free space (vacuum)** at the same speed, $c = 3.00 \times 10^8\,\mathrm{m\,s^{-1}}$. The speed in air is almost the same as the speed in free space. When the waves travel in other mediums (media) their speed is less.

■ The electromagnetic spectrum contains waves of very different wavelengths, sources and properties. The main sections of the electromagnetic spectrum, with their approximate wavelengths are listed in Table 4.1.

> **Expert tip**
>
> Electromagnetic waves from different parts of the spectrum can have very different properties, including their effects on the human body. In order to explain many of these properties we need to consider the *energy* that they each transfer and this requires a knowledge of the *quantum* nature of radiation (see Chapter 7).

Table 4.1

Name	A typical wavelength	Origins	Some common uses
radio waves	10^2 m	electronic circuits/aerials	communications, radio, TV
microwaves	10^{-2} m	electronic circuits/aerials	communications, mobile phones, ovens, radar
infrared (IR)	10^{-5} m	everything emits IR but hotter objects emit *much* more IR than cooler things	lasers, heating, cooking, medical treatments, remote controls
visible light	5×10^{-7} m	very hot objects, light bulbs, the Sun	vision, lighting, lasers
ultraviolet (UV)	10^{-8} m	the Sun, UV lamps	fluorescence
X-rays	10^{-11} m	X-ray tubes	medical diagnosis and treatment, investigating the structure of matter
gamma rays	10^{-13} m	radioactive materials	medical diagnosis and treatment, sterilization of medical equipment

QUESTIONS TO CHECK UNDERSTANDING

16 A mobile phone (such as in Figure 4.13) uses microwaves of frequency 1900 MHz.

   a What is the wavelength of these signals?

   b State one property of microwaves that is different from light.

17 Infrared radiation is emitted by all objects. Suggest two properties of an object which affect the power of the infrared that it emits.

18 Which type of electromagnetic radiation has the highest frequency and what is its origin?

19 Astronomers learn much about the universe from radio waves received on Earth. A common wavelength from hydrogen is 21 cm. What is the frequency of this radio wave in MHz?.

Figure 4.13

The nature of sound waves

Revised ☐

- Sound cannot travel across a vacuum.
- Sound is created when a surface *vibrates* (oscillates) and disturbs the air that surrounds it, producing a series of compressions and rarefactions that travel away from the surface as a longitudinal wave.

> **Key concept**
>
> **Sound** is a longitudinal wave transferred by oscillating molecules, so that it needs a medium through which to travel.

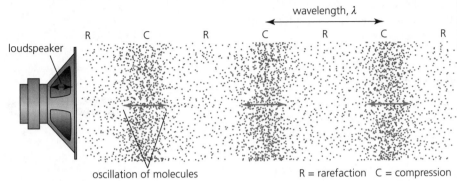

Figure 4.14

- Figure 4.14 represents a sound wave travelling in air away from the surface of a vibrating loudspeaker. A series of compressions and rarefactions are superimposed on the random motions of air molecules.
- As a sound wave passes through air it causes tiny variations in pressure and density.
- The *loudness* of a sound depends on the amplitude of the wave. The *pitch* of a sound depends on the frequency of the wave. The range of hearing of the human ear is approximately 20 Hz to 20 kHz. Higher frequencies are described as **ultrasonic.**
- Figure 4.15 represents two different sound waves. The upper one is a louder and higher-pitched sound than the lower one.
- Sound travels much faster and with less absorption through solids and liquids than through air. This is because the molecules are closer together and have forces between them.

> **Expert tip**
>
> Sound waveforms may be observed using a microphone connected to an oscilloscope.

Investigating the speed of sound experimentally

- Determining the speed of sound requires the use of short, sharp sounds and the measurement of small time intervals, which may involve large uncertainties unless the timing can be done electronically.
- If hand-held stopwatches are used, large distances will be needed for this type of experiment to reduce percentage uncertainties, and/or reflections of sound off large flat surfaces may be useful (echoes).
- Conversely, if the speed of sound in a material is known, the delay involved with echoes can be used to determine distances (echo sounding).

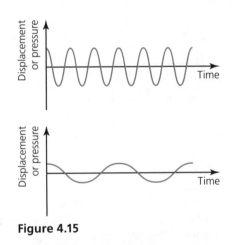

Figure 4.15

QUESTIONS TO CHECK UNDERSTANDING

20 Describe an experiment to determine the speed of sound in air.

21 Four students started stopwatches when they saw a drum being struck a distance of 180 m away. They stopped their watches when they heard the sound. The times were 0.72 s, 0.44 s, 0.58 s and 0.43 s.

   a Why were the times different?

   b Estimate the speed of sound in air from these results.

22 The speed of sound in air at 0 °C is 331 m s⁻¹, but at 40 °C the speed is 353 m s⁻¹.

   a Suggest a reason for the difference.

   b The lowest pitched sound that a student could hear was 15 Hz. What is the wavelength of this sound **i** at 0 °C, **ii** at 40 °C?

23 a Suggest a reason, other than temperature, why the speed of sound in sea water may vary.

   b A sound pulse was transmitted downwards from the bottom of a boat. A strong signal was received back after 0.310 s. How deep was the sea beneath the boat? (Assume that the speed of sound in the water was 1490 m s⁻¹.)

   c Why are short pulses of sound used for this kind of measurement (rather than continuous waves)?

■ Patterns, trends and discrepancies

We know now that sound, light, X-rays, thermal radiation, disturbances from earthquakes (and many other examples) behave in some ways that are very similar. But, historically, it took a long time before the wave nature of all these phenomena was fully recognized. Expertise in one area of knowledge of science may well be transferrable to another. So that, a major aspect of science is looking for such simplicity by detecting patterns and trends in observations.

4.3 Wave characteristics

Revised ☐

Essential idea: All waves can be described by the same sets of mathematical ideas. Detailed knowledge of one area leads to the possibility of prediction in another.

Wavefronts and rays

Revised ☐

- On paper we represent waves with lines called *wavefronts*. For example a line joining points along the crest of a wave.
- Adjacent wavefronts are one wavelength apart.
- A line showing the direction in which wavefronts are moving is called a *ray*. Rays are always perpendicular to wavefronts.
- A ray coming into a surface/boundary is called an *incident* ray.

■ Sketching and interpreting diagrams involving wavefronts and rays

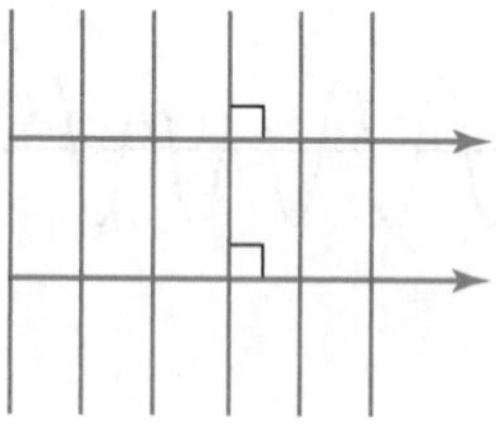

> **Key concepts**
>
> A **wavefront** is a line joining points next to each other that are moving in phase.
>
> Lines showing the direction in which wavefronts are moving are called **rays**.

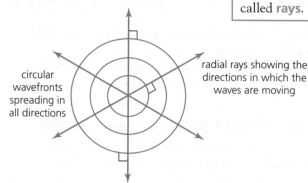

parallel rays

plane wave fronts

Figure 4.16

circular wavefronts spreading in all directions

radial rays showing the directions in which the waves are moving

Figure 4.17

- If wavefronts are straight and parallel to each other they can be described as (idealized) **plane waves**. The movement of plane waves is represented by parallel rays. See Figure 4.16.
- *Circular wavefronts* are represented by *radial* rays spreading out from the source of the waves. See Figure 4.17.
- School laboratories often use small *ripple tanks* to demonstrate the behaviour of wavefronts on the surface of water.
- The concepts of wavefronts and rays are used extensively in explaining the behaviour of all waves (see later).

Amplitude and intensity

Revised ☐

- Because we are usually concerned with waves that are *continuously* emitted or received, it is unusual to discuss the energy of a wave. We are much more likely to be concerned with wave *power* (energy transferred/time).
- Typically, the power of waves spreads out in three dimensions, so that the concept of power passing through, or arriving at, unit area becomes very useful. This is called the intensity, *I*, of a wave and it has the unit W m^{-2}.
- The property of a wave itself that is related to intensity is wave *amplitude*. For example, a wave with 3 × the amplitude of another similar wave is transferring $3^2 = 9$ × the power per square metre (intensity).

> **Key concept**
>
> **Intensity**, *I*, is defined as the power passing perpendicularly through unit area, $I = \dfrac{P}{A}$.
>
> The intensity of a wave is proportional to the amplitude squared: $I \propto A^2$.

- The relationship between intensity and amplitude can be expressed as $\frac{I}{A^2} =$ constant.

Expert tip

If waves strike a surface at an angle of less than 90°, they will be spread over a greater area and the *received* intensity will be reduced.

Inverse square law

Revised ▢

- It is common to want to know how the intensity of a wave changes as it moves away from a source. For example, how the intensity of light changes with distance from the lamp emitting it, or how sound level varies with distance from a loudspeaker.
- For simplicity, we will assume that the waves come from a *point source* and spread equally in all directions (in three dimensions) without any absorption (energy dissipation).

Expert tip

In reality, many waves are *not* emitted equally in all directions. This may be because of the size or shape of their source, or because some kind of reflectors are used to send the waves in a particular direction, or because the waves are not travelling in three dimensions. Nevertheless, the inverse square law is a very important starting point in an understanding of wave intensities.

- Figure 4.18 shows three imaginary spheres around a point source, P, of waves at distances x, $2x$ and $3x$. The surface area of any sphere is $4\pi r^2$, so the surface areas of these three spheres are $4\pi x^2$, $16\pi x^2$ and $36\pi x^2$. More simply, the areas are in the ratio $1 : 2^2 : 3^2$. If the intensity of the waves passing through the first sphere is I, the intensity at the second is $\frac{I}{2^2}$ and at the third the intensity is $\frac{I}{3^2}$.

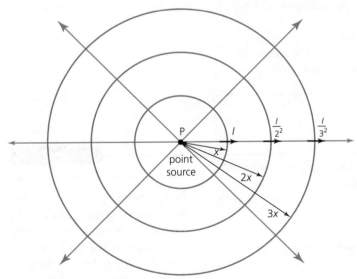

Figure 4.18

> **Key concept**
>
> The **inverse square law**: for waves spreading out equally in three dimensions from a point source without any loss of energy, their intensity, I, is inversely proportional to the distance from the source, x, squared: $I \propto x^{-2}$.

- The relationship-between intensity and distance can be expressed as $Ix^2 =$ constant.
- Figure 4.19 shows a simplified numerical example of a graph of intensity–distance for an inverse square law relationship (arbitrary units).
- Figure 4.20 shows the same data used to produce a straight-line graph.

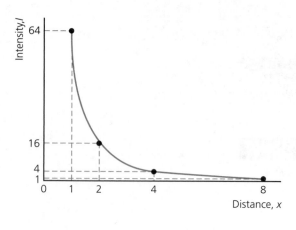

Figure 4.19

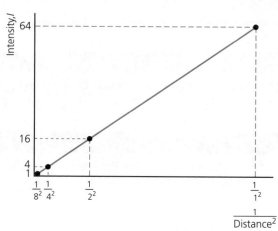

Figure 4.20

■ Solving problems involving amplitude, intensity and the inverse square law

QUESTIONS TO CHECK UNDERSTANDING

24 The intensity of radiation falling normally on a solar heating panel was 520 W m^{-2}. If the area of the panel was 4.2 m^2, how much energy arrived in one hour?

25 a If the amplitude of a wave increases by 25%, what is the corresponding increase in intensity?

  b If the intensity of a wave drops to 10% because of energy dissipation, what has happened to its amplitude?

26 An electrical generator designed to transfer energy from sea waves produces an output power of 10 kW when the average wave amplitude is 2.0 m. Estimate the output when the wave amplitude falls to 1.5 m.

27 The distance of Mars from the Sun is about 1.5 × the distance of the Earth from the Sun. How much greater is the intensity of solar radiation arriving at Earth, compared to Mars?

28 Give two reasons why the inverse square law cannot be used to accurately predict the sound levels at varying distances from a television set in a living room.

29 A man is looking at a map at night using only the light from a street lamp 30 m away. Assuming that the lamp emits equally in all directions, how far would he have to walk towards the lamp in order for the light intensity to double?

Superposition

Revised ☐

- When two, or more, waves (or pulses) arrive at the same point, the result can be predicted by adding (superposing) the waves together.
- The superposition of waves is important because only waves pass through each other and combine in this way.

■ Sketching and interpreting the superposition of pulses and waves

- Most commonly, examples of superposition involve wavefronts from two sources, like A and B in Figure 4.21, crossing and passing through each other at any point (P).
- Figure 4.22 shows possible displacement–time graphs for the waves from A and B, together with the resultant wave formed at P.

> **Key concept**
>
> **Principle of superposition:** the overall displacement at any point at any time will be the vector sum of all the individual wave displacements.

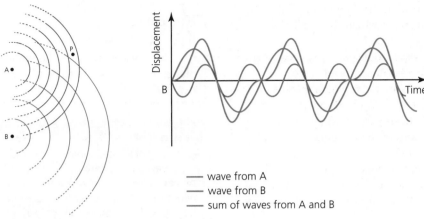

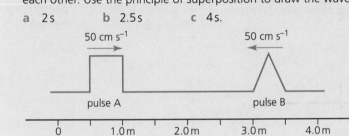

—— wave from A
—— wave from B
—— sum of waves from A and B

Figure 4.21   **Figure 4.22**

- In this general example we have considered waves of different frequency and amplitude, but the most important application of the superposition principle involves waves of the same frequency and similar amplitudes: see the topic of *interference* in Section 4.4.

QUESTIONS TO CHECK UNDERSTANDING

30 Figure 4.23 shows two idealized pulses on a stretched rope moving towards each other. Use the principle of superposition to draw the waves after

    a 2 s   **b** 2.5 s   **c** 4 s.

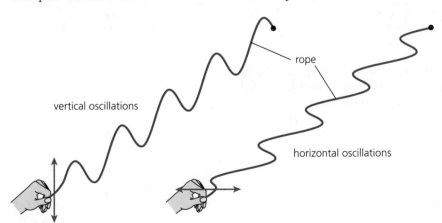

Figure 4.23

31 **a** On the same displacement–time axes (0–2 s), carefully sketch two sinusoidal waveforms to represent:

    i a wave of amplitude 3.0 cm and frequency 2 Hz

    ii a wave of amplitude 2.0 cm and frequency 1 Hz.

  b Use the principle of superposition to sketch how these two waves would combine if they met at a point.

Polarization

Revised ☐

- Consider sending transverse waves along a stretched rope, as in Figure 4.24. In principle, you could create the transverse waves by shaking your hand in *any* direction which was perpendicular to the rope. However, in practice, you would probably only shake the rope in one plane. Figure 4.24 shows two examples: oscillations in the vertical and horizontal planes.

vertical oscillations

rope

horizontal oscillations

Figure 4.24

- When the oscillations of a transverse wave all occur in the same plane, the wave is described as being *polarized*. The plane in which the oscillations occur is called the **plane of polarization**.
- Note that it is impossible for longitudinal waves to be polarized because their oscillations are always in the same direction as energy transfer.
- The most important examples of polarization are those which involve electromagnetic waves. Refer back to Figure 4.12. Electromagnetic waves consist of oscillating electric and magnetic fields perpendicular to each other. Because they are transverse waves, both of these oscillations are also perpendicular to the direction of energy transfer.
- When discussing oscillations within an electromagnetic wave we usually concentrate our attention on the variations of the electric field vector. Figure 4.12 only shows one direction of oscillation for this vector (horizontal), so it is representing a polarized wave.
- Most electromagnetic waves are emitted from their sources in a random and uncontrolled way, so electromagnetic waves are usually *unpolarized*. That is, they contain a mixture of all possible planes of polarization.

> **Key concept**
>
> If the oscillations transferring a transverse wave are all in the same plane, the wave is said to be **plane polarized**.

> **Expert tip**
>
> Artificially produced radio waves and microwaves are polarized because of the controlled way in which they are made: the electric currents producing them flow (through aerials) only in certain directions.

■ Describing methods of polarization

- Light is normally unpolarized because oscillations occur in all possible directions. If we wish to produce plane polarized light from normal light we need a method which removes the oscillations in other planes.
- A polarizing filter absorbs the energy of oscillations in all planes but the plane of polarization. See Figure 4.25.

> **Key concept**
>
> Unpolarized light can be polarized by passing it through a **polarizing filter** (also called a **polarizer**).

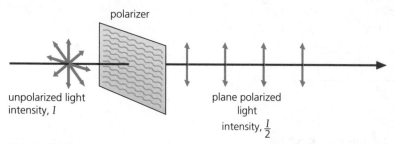

Figure 4.25

- If the polarized light is then passed through a second polarizing filter (often called an **analyser**), the intensity of the transmitted light will be zero if the filters are *crossed*, and remains almost unaltered if the filters are aligned. See Figure 4.26.

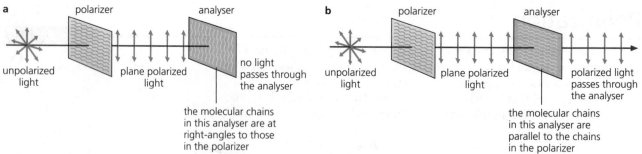

Figure 4.26

- Figure 4.27 shows the effect of looking through two polarizing filters. In Figure 4.27a the filters are aligned, in Figure 4.27c the filters are *crossed* and no light can pass through, and in Figure 4.27b some light is absorbed (depending upon the angle, θ).

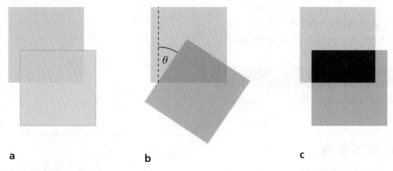

a   **b**   **c**

Figure 4.27

Expert tip

A polarizing filter contains long chain molecules parallel to each other in one direction. If an electric field vector within an electromagnetic wave is aligned with the molecules, energy will be absorbed. The plane of polarization produced is perpendicular to the plane of the molecules.

- Unpolarized light will also become partially polarized when it is reflected off an insulator such as glass or water. The amount of polarization depends on the material involved and the angle of incidence. See Figure 4.29 below.
- Light reflected off various surfaces can be viewed through a rotated polarizing filter (analyser). Any change in received intensity indicates that the reflected light is polarized to some extent.

Sketching and interpreting diagrams illustrating polarized, reflected and transmitted beams

Key concept

Light may become polarized to some extent when it reflects off (or passes through) certain insulators.

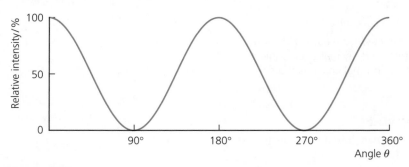

Figure 4.28

- Figure 4.28 shows how the transmitted intensity through two polarizing filters changes as one of them is rotated. θ is the angle between the transmission axes of the two filters. If $\theta = 0°$ or $180°$ the filters are aligned, and if $\theta = 90°$ or $270°$ the filters are crossed. This behaviour is described mathematically by *Malus's law* (see below).
- When unpolarized light is incident from air upon a transparent medium (like glass or water) some of the light is reflected and some of the light is transmitted as a refracted beam (see Section 4.4 for details about refraction). The reflected beam will always be polarized to some extent, with the electric field vectors parallel to the surface. The transmitted beam must contain the light which was not reflected, so it too must be polarized to some extent, but in a plane perpendicular to the surface. See Figure 4.29a.

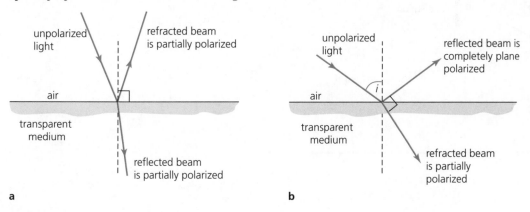

a   **b**

Figure 4.29

■ At a certain angle of incidence for light passing from air into a particular medium, the reflected beam is *completely* plane polarized. This occurs when the reflected and transmitted wavefronts (and rays) are perpendicular to each other. See Figure 4.29b.

Expert tip

The angle of incidence for which the reflected beam is totally polarized is called the Brewster angle.

■ Solving problems involving Malus's law

■ *Malus's law* mathematically describes the variation of intensity as plane polarized light passes through a rotating polarizing filter: $I = I_0 \cos^2 \theta$, where I_0 is the incident intensity, I is the transmitted intensity and θ is the angle between the plane of polarization of the incident light and the axis of transmission for the filter. (See Figure 4.28 for graphical representation.)

> **Key concept**
>
> **Malus's Law** can be used to predict the intensity of light which emerges from an analyzing filter: $I = I_0 \cos^2 \theta$.

QUESTIONS TO CHECK UNDERSTANDING

32 a Explain why light can be polarized but sound cannot.

b Explain why light waves are *not* usually polarized.

33 Unpolarized light of intensity I is passed through two polarizing filters. If the axes of transmission for the two filters are at 40° to each other,

a what is the light intensity between the filters?

b What light intensity is transmitted by the second filter?

34 a A student discovers that a ray of light reflected from a glass block is completely plane polarized when the refracted ray makes an angle of 33° with the normal. At what angle to the normal (in air) is the ray incident upon the block?

b What is the refractive index of the glass?

c In which plane is the reflected wave polarized?

d Describe how the student could have determined that the reflected ray was completely polarized.

35 Explain why photographers sometimes like to use polarizing filters in front of their camera lenses.

36 When polarized light is passed through some chemical solutions the plane of polarization is changed (rotated). See Figure 4.30. If the intensity is I before the solution is placed in the beam and $0.9I$ after the sugar is placed in the beam, estimate the angle of rotation.

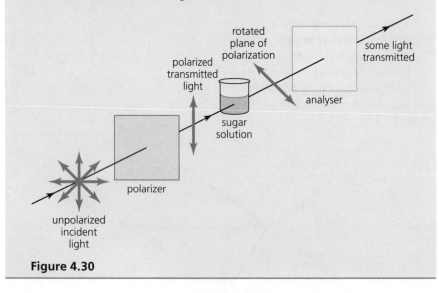

Figure 4.30

■ Imagination

The importance of *imagination* in scientific and technological advances is often undervalued compared with other human endeavours. It may seem as if scientific knowledge is just out there awaiting discovery in due course and, when we look back with today's knowledge, the theories of great scientists of the past may seem unsurprising, almost obvious. But such hindsight belittles the work of great thinkers who used their imagination to think in ways different from their contemporaries.

Imagination flourishes when stimulated by interesting and unexplained observations, such as the polarization of light passed through certain crystals, which was first seen over 1300 years ago.

4.4 Wave behaviour

Revised ☐

Essential idea: Waves interact with media and each other in a number of ways that can be unexpected and useful.

■ *All* waves reflect, refract, diffract and interfere under suitable conditions. We will consider each of these properties in Section 4.4.

Reflection

Revised ☐

■ The **reflection** of light is a topic usually well covered in early physics classes using the concept of light *rays*. To summarize: when a light ray strikes a plane (straight) surface, the angle of incidence, *i*, is equal to the angle of reflection, *r*, and they are in the same plane (**law of reflection**). See the *ray* diagram in Figure 4.31. Note that the angles are measured to a '**normal**': an imaginary line perpendicular to the surface.

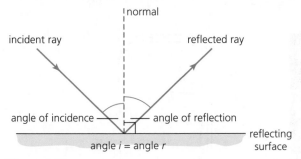

Figure 4.31

■ Sketching and interpreting incident and reflected waves at boundaries between media

■ However, remember that rays are only lines showing the direction in which wavefronts are travelling, so that we also need to be able to represent reflection by *wave* diagrams. See Figure 4.32.

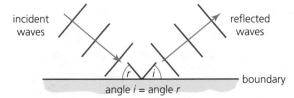

Figure 4.32

> **Key concept**
>
> When wavefronts reflect off a plane surface the angle of incidence equals the angle of reflection.

■ Reflection can occur when *any* wave meets a boundary between two different materials. The reflection of light is just the most common everyday example.

Waves are reflected with a phase change of half a wavelength (π radians) from a boundary with a medium into which they cannot pass, or in which they travel slower. (This information is needed later.) Consider the one-dimensional example shown in Figure 4.33.

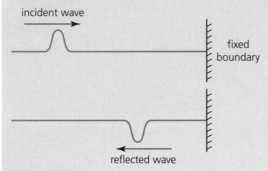

Figure 4.33

■ Solving problems involving reflection at a plane interface

■ We are only concerned here with the reflection of plane waves from plane surfaces.

We can see objects because their surfaces reflect light into our eyes. This diffuse and irregular kind of reflection (scattering) of light is more difficult to analyse than the plane waves and surfaces discussed here.

QUESTIONS TO CHECK UNDERSTANDING

37 The structure shown in Figure 4.34 has been built along the side of a motorway. Suggest its purpose.

Figure 4.34

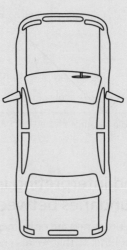

Figure 4.35

38 Figure 4.35 shows the location of wing mirrors on the side of a car. Make a quick copy of the diagram and use it to explain why the mirrors used are *not* plane mirrors.

39 Figure 4.36 shows some wavefronts striking a plane mirror. The dotted lines shows where the wavefronts would be if the mirror was not in their path. Copy the diagram and show the actual positions of these four reflected wavefronts.

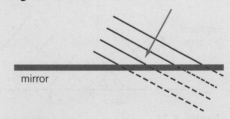

Figure 4.36

Refraction

- In general, when waves are incident upon a boundary between two different media, apart from being reflected, some of the wave energy may also be transmitted into the second medium. The wave speeds in the two media will probably be different. However, the wave frequency cannot change.
- If, for example, the wave speed decreases in the second medium, the wavelength must also decrease. This is because $c = f\lambda$ and f is constant.
- When waves are transmitted obliquely (not perpendicularly) into a different medium the change of speed results in a change of direction which is called **refraction**.

■ Sketching and interpreting incident and transmitted waves at boundaries between media

- Figure 4.37 shows waves being refracted as they pass into a medium where they travel slower (for light such a material is often described as an *optically denser medium*). Under these circumstances the ray is refracted *towards the normal*.

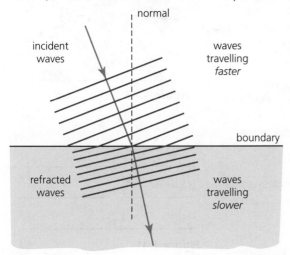

Figure 4.37

- The amount of refraction depends on how much the wave speed has changed.
- The *refractive index*, n, of a medium is a constant which indicates how much the medium will refract light that enters into it *from a vacuum*. The values of refractive indices depend on how much the speed of light changes as light enters different mediums from vacuum.
- For example, the refractive index of a certain kind of glass might be 1.5 because the speed of light in that glass is 2.0×10^8 m s^{-1}. The refractive index of air is 1.0.

■ Snell's law

- This law links the angles of incidence and refraction to the speeds of the waves. Consider Figure 4.38.

> **Key concept**
>
> When plane waves enter a medium where they travel slower, they are refracted towards the normal (and vice versa).

> **Key concept**
>
> **Refractive index** of a medium, n = speed of waves in vacuum (3.0×10^8 m s^{-1})/speed of waves in the medium.

> **Key concept**
>
> Snell's law: $\dfrac{n_1}{n_2} = \dfrac{\sin\theta_2}{\sin\theta_1} = \dfrac{v_2}{v_1}$

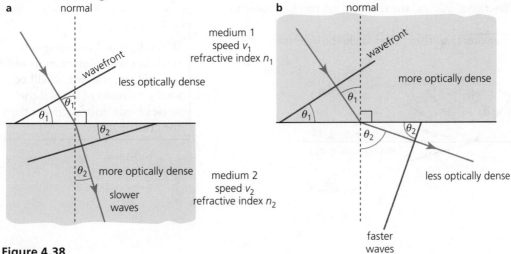

Figure 4.38

■ Determining refractive index experimentally

■ Look at Figure 4.38a again. If medium 1 is air and medium 2 is the material under investigation, then n_2, the refractive index of medium 2, can be found from $n_2 = \dfrac{\sin \theta_1}{\sin \theta_2}$. The refractive index of a medium can be found from measurement of the angles of incidence and refraction. This is usually done by tracing rays through a parallel-sided glass block. See Figure 4.39.

■ Solving problems involving Snell's law

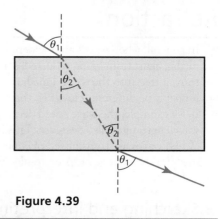

Figure 4.39

QUESTIONS TO CHECK UNDERSTANDING

40 Describe an experiment to determine the refractive index of glass.

41 Light travelling *in* water strikes the surface from below. Some waves are reflected and some are refracted. Draw a sketch to show how four separate wavefronts are affected by reflection and refraction at the surface.

42 Light enters a flat surface of a plastic block at an angle of incidence of 38°. The plastic has a refractive index of 1.41.

   a   What is the speed of light in the plastic?

   b   What is the angle of refraction in the plastic?

43 (Monochromatic) light enters a triangular glass prism at an angle of incidence of 42° as shown in Figure 4.40.

   a   If the angle of refraction was 26°, what was the refractive index of the glass?

   b   Determine the angle θ.

44 Figure 4.41 shows a ray of light passing through parallel layers of glass and a transparent plastic.

   a   Does the light travel faster in the glass or the plastic?

   b   Determine the angle of incidence, *i*.

45 Refraction occurs when wave speeds change. Suggest how light and sound waves may be affected by passing into the air above a hot surface.

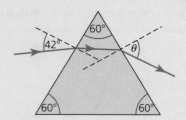

Figure 4.40

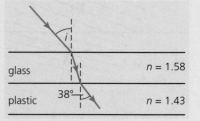

Figure 4.41

Critical angle and total internal reflection

Revised ☐

■ When light travelling in an optically denser medium strikes a boundary with a less optically dense medium (lower refractive index) it may be **totally internally reflected**, so that no light is transmitted out of the medium. A common example is light travelling in glass surrounded by air.

■ Figure 4.42 illustrates light rays in an optically denser medium with different angles of incidence on the boundary. At a certain angle, called the **critical angle**, the angle of refraction is 90° and the refracted ray travels parallel to the boundary.

■ For angles of incidence greater than the critical angle the light is totally internally reflected.

> **Key concept**
>
> When light in incident upon a boundary with a medium in which it would travel slower, it will be *totally internally reflected* if the angle of incidence is greater than the critical angle.

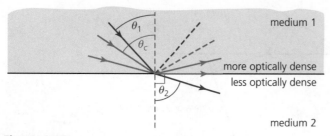

Figure 4.42

■ Solving problems involving critical angle and total internal reflection

■ When the angle of incidence is the critical angle, c, the angle of refraction is 90° and its sine is 1. So that Snell's law reduces to: $\sin c = \dfrac{n_2}{n_1}$, where n_1 is the refractive index of the optically denser medium.

QUESTIONS TO CHECK UNDERSTANDING

46 The critical angle for light within a certain plastic surrounded by air is 39°.
   a   What is the refractive index of the plastic?
   b   If the plastic was placed in water, would its critical angle increase, decrease or stay the same?

47 The speed of light in water is $2.25 \times 10^8 \, \text{m s}^{-1}$ and in a certain kind of glass it is $1.95 \times 10^8 \, \text{m s}^{-1}$.
   a   What is the critical angle between these two media?
   b   In which medium would light have to be travelling in order to be totally internally reflected?

48 Draw a curved 'light pipe' transmitting a totally internally reflected ray of light.

Figure 4.43

Dispersion of light by a prism

Revised

■ Although all electromagnetic waves have the same speed in air, different colours of light have very slightly different speeds in other media. This means that they also have slightly different refractive indices.
■ As a result, different colours will follow slightly different paths through a *prism* and this results in the **dispersion** of white light into a spectrum, as shown in Figure 4.44.

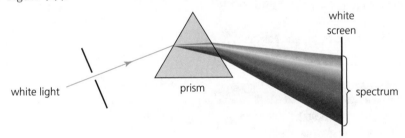

Figure 4.44

Diffraction through a single slit and around objects

Revised

■ **Diffraction** is the spreading of waves as they pass through gaps and around obstacles. All types of waves can diffract under suitable circumstances. Waves diffracting on the surface of water is a common sight.
■ Figure 4.45 shows some examples of the diffraction of plane waves. These patterns are commonly demonstrated on ripple tanks in school laboratories.

Key concept

Diffraction is most significant when the gap or obstacle size is similar in magnitude to the wavelength, see Figure 4.45

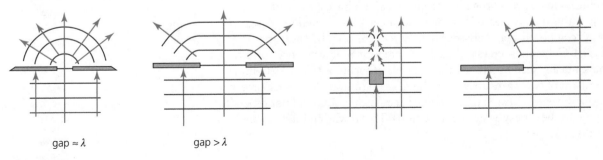

gap ≈ λ   gap > λ

Figure 4.45

- Apart from water waves, the diffraction of sound is also easy to demonstrate. Sound can 'go around corners' because it has wavelengths that are similar to the size of common objects. Of the waves of electromagnetic spectrum, the diffraction of microwaves (with wavelengths of a few centimetres) through a slit between metal sheets is a common laboratory demonstration.

■ Qualitatively describing the diffraction pattern formed when plane waves are incident normally on a single slit

- The diffraction of light is less easy to observe because light has a very small wavelength ($\approx 5 \times 10^{-7}$ m). However, a diffraction pattern can easily be produced by passing laser light through a very narrow gap. See Figure 4.46.
- The central band is broadest and brightest. To begin to understand the pattern we need to realize that even the narrowest slit is many times wider than the wavelength of light.

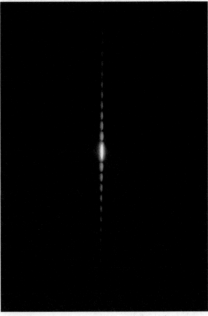

Figure 4.46

QUESTIONS TO CHECK UNDERSTANDING

49 When sound is received from a source which is about 100 m or more away, explain why low-frequency sounds will probably be easier to hear than high-frequency sounds.

50 A mobile phone company uses a frequency of 1900 MHz. Estimate the size of an emitting aerial which will ensure that the waves spread out well from their source in the tower.

51 Diffraction is to be demonstrated in the laboratory by observing how waves on water in a ripple tank pass through a gap of width 1 cm. If the speed of the waves is 20 cm s^{-1}, what frequency would be needed for the greatest diffraction effects?

52 a How would the diffraction pattern shown in Figure 4.46 change if the slit was made narrower?

  b Suggest how the pattern would change if the monochromatic light was changed to a white light source.

Expert tip

An explanation of the single-slit diffraction pattern of light requires a knowledge of *interference* (see next sub-section). If the narrow slit was, for example, 100× the wavelength of light, then we may imagine that there are 100 sources of *coherent* circular *wavelets* diffracting away from the slit. These wavelets then *interfere* to produce the *'diffraction'* pattern. There is a fuller explanation of this in Chapter 9 for HL students.

Interference patterns

Revised ▢

- Under certain circumstances, when wavefronts spread away from two sources and then pass through each other, they can combine to produce a pattern which stays constant in time. This is called an **interference pattern**.
- Figure 4.47 shows wavefronts spreading away from two sources S$_1$ and S$_2$. Interference effects may be observed in the shaded area.
- In this section we will discuss only the relatively simple situations in which two sources emit waves of the same frequency which have a *constant phase difference* (this usually means that the waves are always emitted in phase, or always completely out of phase). Such sources are described as being **coherent**. Coherent sources are needed to produce an interference pattern.
- We have already seen how to use the principle of superposition to determine the resultant of two waves arriving at the *same* point from different sources. However, waves from two sources received at *different* points will not generally superpose in the same way (at the same time). This is why an interference pattern is produced.

- If waves arrive *in phase* at a point, superposition will produce a wave of greater amplitude (double the amplitude if the waves were identical). This is called **constructive interference**. See Figure 4.48a. Figure 4.48b shows the effect of waves arriving *completely out of phase*: the resulting amplitude will be much reduced and would be zero if the waves had exactly equal amplitudes. This is called **destructive interference**. At most places the waves will *not* arrive perfectly in phase or perfectly out of phase, but the waves still interfere.

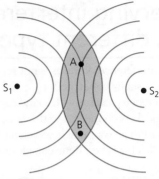

Figure 4.47

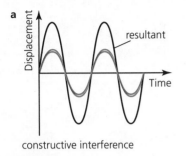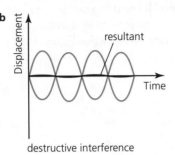

Figure 4.48

Path difference

- In order to predict or explain the shape of an interference pattern we need to compare the distances travelled by wavefronts from the two sources.
- Consider again Figure 4.47. The result of superposition at, for example, point A and point B will be different because the distances $(S_1A–S_2A)$ and $(S_1B–S_2B)$ are different. The difference between these two distances is known as their *path difference*
- Figure 4.49 shows the path differences involved with five maxima.

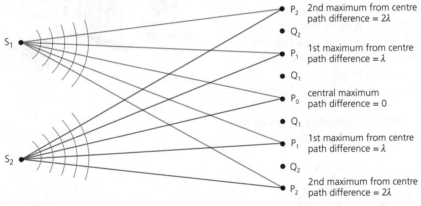

P_2 2nd maximum from centre path difference = 2λ

Q_2

P_1 1st maximum from centre path difference = λ

Q_1

P_0 central maximum path difference = 0

Q_1

P_1 1st maximum from centre path difference = λ

Q_2

P_2 2nd maximum from centre path difference = 2λ

Figure 4.49

- The shape of the interference pattern produced by these conditions (in two dimensions) is shown in Figure 4.50. This is often demonstrated on a ripple tank. The line marked C_0 joins places where the path difference is zero, the line marked C_1 joins places where the path difference is one wavelength, the line marked $D_{\frac{1}{2}}$ joins places where the path difference is half a wavelength, the line marked $D_{\frac{3}{2}}$ joins places where the path difference is one and a half wavelengths, etc.

> **Common mistake**
>
> Two waves combining to give no waves may seem to contradict the principle of conservation of energy, but in other places the wave amplitude is doubled, which suggests four times the intensity (Section 4.3: wave intensity is proportional to amplitude squared). Overall, energy is conserved.

> **Key concept**
>
> **Path difference** is the difference in distance travelled by two waves from their sources to a given point.
>
> Constructive interference occurs when path difference = a whole number of wavelengths ($n\lambda$).
>
> Destructive interference occurs when path difference = an odd number of half wavelengths ($(n + \frac{1}{2})\lambda$).

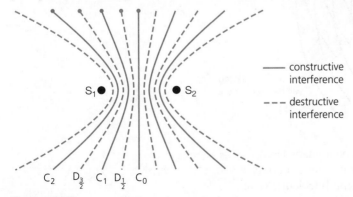

—— constructive interference

--- destructive interference

C_2 $D_{\frac{3}{2}}$ C_1 $D_{\frac{1}{2}}$ C_0

Figure 4.50

Observing interference patterns with different types of wave

- The interference of waves is not an everyday observation because wave sources are *not* usually coherent.
- Interference patterns with *sound* can easily be produced by using two loudspeakers connected to the same signal generator. If the frequency is adjusted to give a suitable wavelength, we can detect the pattern by moving around. This is best done outside, away from reflecting surfaces. See Figure 4.51.

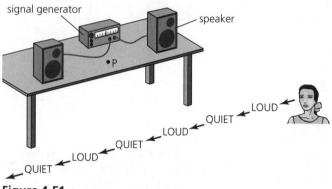

Figure 4.51

■ Double slit interference

- The interference of *microwaves* ($\lambda \approx 3$ cm) can be demonstrated by effectively creating two coherent sources by putting two 3 cm slits in front of a single microwave source. Because the slit width is about the same as the wavelength, the diffracted wavefronts emerging from the slits in a horizontal plane will be semi-circular. These waves will then overlap and interfere. See Figure 4.52.

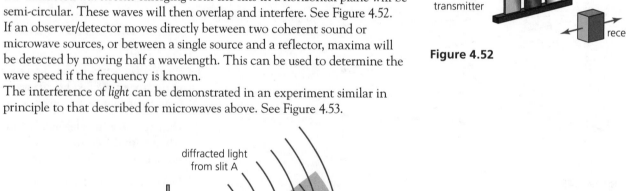

Figure 4.52

- If an observer/detector moves directly between two coherent sound or microwave sources, or between a single source and a reflector, maxima will be detected by moving half a wavelength. This can be used to determine the wave speed if the frequency is known.
- The interference of *light* can be demonstrated in an experiment similar in principle to that described for microwaves above. See Figure 4.53.

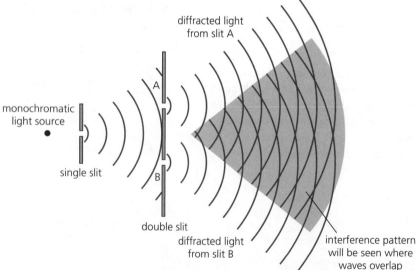

Figure 4.53

- Figure 4.54 shows the experimental arrangement: monochromatic laser light provides an intense and suitable light source (but it is not essential). Monochromatic means consisting of only one wavelength (colour). More details of this experiment are provided in Section 9.3 (HL).

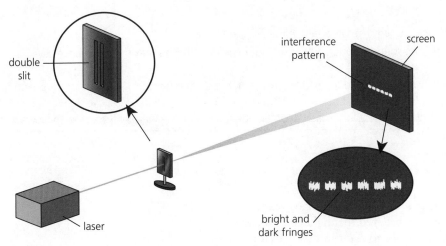

Figure 4.54

■ A pattern of (almost) equally spaced bright and dark '*fringes*' is seen in the centre of a screen. The pattern can *only* be explained in terms of wave superposition. This experiment has great historical significance because, when the interference of light was observed (by Young), it provided the first experimental proof that light travelled as waves. The wavelength of light can be determined from suitable measurements. See below.

Quantitatively describing double-slit interference intensity patterns

■ Figure 4.55 shows the basic geometry of the experiment. An unknown wavelength can be determined by measuring the average separation of the centres of the fringes, *s*, and using the following equation:

$s = \dfrac{\lambda D}{d}$, where D is the slit-to-screen distance and d is the distance between the centres of the slits.

> **Common mistake**
>
> Single-slit diffraction patterns of light are easily confused with double-slit interference patterns. But, in a single slit diffraction pattern the central band is brighter and wider than the others.

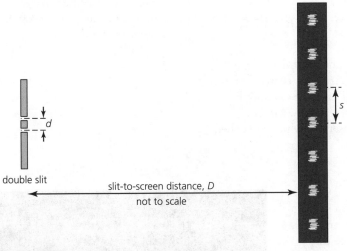

double slit

slit-to-screen distance, *D*

not to scale

Figure 4.55

QUESTIONS TO CHECK UNDERSTANDING

53 a Explain why two separate light sources cannot be used to produce an interference pattern.

b Why are monochromatic light sources often preferred for interference experiments?

c In a double-slits light experiment the two slits were half a millimetre apart and they produced a fringe pattern in which the distance across 10 fringes was measured to be 1.9 cm on a screen which was placed 1.98 m from the slits. Calculate the wavelength used.

54 A student walks slowly between two loudspeakers which are facing each other and both emitting sound of wavelength 1.20 m.

a What is the shortest distance that a student has to walk to move between two positions of minimum intensity?

b Why are experiments like this better done outdoors?

55 Consider Figures 4.49 and 4.51.

   a Explain why the sound heard by the student rises and falls in intensity in Figure 4.51.

   b *Estimate* an approximate value for the sound frequency that would be needed for the maxima to be heard every metre along the path shown.

   c Is it possible for the sound intensity to be reduced to zero at any location (assume that both speakers produce sound of the same intensity)?

56 Explain how the apparatus shown in Figure 4.52 could be used to determine a value for the wavelength of the microwaves.

■ Competing theories

The true nature of light was a major issue in science for many, many years: did light consist of waves or particles? Newton believed that light consisted of particles, however the interference of light seemed to confirm that light was wave-like in nature, but that theory was also shown to be imperfect by Einstein's photon explanation of the photoelectric effect (see Chapters 7 and 11).

4.5 Standing waves

Revised ▢

Essential idea: When travelling waves meet they can superpose to form standing waves in which energy may not be transferred.

The nature of standing waves

Revised ▢

■ In Section 4.2 we discussed *travelling waves*, which transfer energy away from a source. In this section we will introduce a different type of wave: **standing waves**, which are sometimes called stationary waves. Standing waves are most commonly observed when waves are 'trapped' in a system and reflect back on themselves. The wave pattern stays in the same place.

■ Describing the nature and formation of standing waves in terms of superposition

■ The principle of superposition can be used to determine what happens if two waves of the same frequency and amplitude pass through each other travelling at the same speed in opposite directions. This most commonly occurs when waves are reflected back upon themselves at some kind of *boundary*. At certain frequencies patterns may be observed, which are called standing waves.

■ Nodes and antinodes

■ In principle, standing waves can occur with any type of wave, but the most commonly visualized example is shown in the photos of Figure 4.56, which compares four standing waves of different frequencies on the same stretched string. Note that in each of these four examples, there are places where the string is stationary. These points are called **nodes**. The two waves arriving in opposite directions at a node are always exactly out of phase.

■ The distance between adjacent nodes (or antinodes) is half a wavelength.

■ There are also places called **antinodes**, where the string is oscillating with its maximum amplitude. The two waves arriving in opposite directions at an antinode are always exactly in phase.

■ It should be noted that, apart from at the nodes, the string is oscillating all the time and a pattern is seen only because the string has completed many oscillations in the time during which the photo was taken.

> **Key concept**
> When waves are reflected back on themselves in confined spaces, at certain frequencies, superposition can result in standing waves.

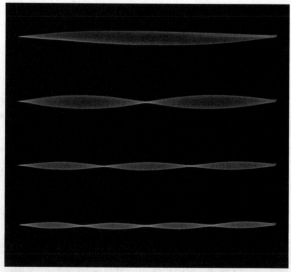

Figure 4.56

Distinguishing between standing and travelling waves

Revised ☐

■ Travelling waves (as discussed in Section 4.2) transfer energy away from a source, but there is no transfer of energy by standing waves. Table 4.2 summarizes the differences between these two types of wave.

Table 4.2

	Standing waves	Travelling waves
Wave pattern	stationary pattern of nodes and antinodes	progressive/travelling
Energy transfer	no energy is transferred	energy is transferred in the direction of wave travel
Amplitude (assuming no energy dissipation)	amplitude at any one place is constant but it varies with position between nodes; maximum amplitude at antinodes, zero amplitude at nodes	all oscillations have the same amplitude
Phase	all oscillations between adjacent nodes are in phase	oscillations one wavelength apart are in phase; oscillations between are not in phase
Frequency	all oscillations have the same frequency	all oscillations have the same frequency
Wavelength	twice the distance between adjacent nodes	shortest distance between points in phase

Observing, sketching and interpreting standing wave patterns in strings and pipes

Revised ☐

■ The IB Physics course concentrates on standing waves of two particular kinds: *transverse waves on stretched strings and longitudinal sound waves in pipes* containing air (air columns).

■ It is possible for a particular system to vibrate with standing waves of different wavelengths and frequencies: these are called different **modes of vibration** or **harmonics**. The lowest frequency (longest wavelength) is called the **first harmonic**, f_0. Other harmonics are multiples of f_0, but they may not all be possible on a particular system. This depends on the *boundary conditions* of the particular system. See below.

■ If a stretched string is encouraged to oscillate (by plucking, for example), it will usually do so predominantly in its first harmonic mode. But strings and air columns can also be made to oscillate in other modes if they are disturbed by external oscillations of suitable frequencies.

■ Figure 4.57 shows the first four harmonics of a string that is fixed at both ends (the most common arrangement). Compare this to Figure 4.56.

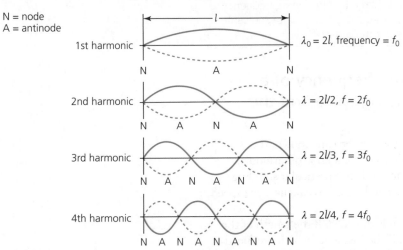

N = node
A = antinode

1st harmonic — N   A   N — $\lambda_0 = 2l$, frequency = f_0

2nd harmonic — N A N A N — $\lambda = 2l/2$, $f = 2f_0$

3rd harmonic — N A N A N A N — $\lambda = 2l/3$, $f = 3f_0$

4th harmonic — N A N A N A N A N — $\lambda = 2l/4$, $f = 4f_0$

Figure 4.57

■ Boundary conditions

■ For stretched strings, the boundary conditions are that there will be nodes at fixed boundaries and antinodes at free boundaries. Figure 4.57 shows the most common example: two fixed ends. Note that all harmonics are possible on this system.

■ For pipes (air columns), the boundary conditions are that there will be nodes at closed boundaries and antinodes at open boundaries. As an example, Figure 4.58 shows the first three harmonics in a pipe which is open at one end and closed at the other. Note that even numbered harmonics are not possible on this particular system.

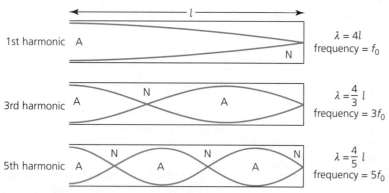

Figure 4.58

QUESTIONS TO CHECK UNDERSTANDING

57 Figure 4.59 shows a demonstration of standing waves in air in a long tube. To begin with a powder was spread along the tube, but when a certain frequency was produced by the loudspeaker the powder moved into piles as shown.

   a What do the red lines represent?

   b Explain why the powder moves into piles marked N.

   c Which harmonic is shown in the figure?

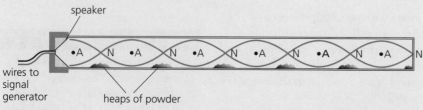

Figure 4.59

58 Sketch standing wave patterns for the first three possible harmonics for waves on a stretched string which was fixed at one end and free at the other. Make a note of the wavelength in terms of the length of the string besides each pattern.

59 An organ pipe which was open at both ends had a first harmonic frequency of 80 Hz. What would be the first harmonic frequency of a pipe of the same length which was open at one end a closed at the other?

■ Solving problems involving the frequency of a harmonic, length of the standing wave and the speed of the wave

■ If the length of the system and its boundary conditions are known, the wavelength, λ, of a standing wave can be deduced for a known harmonic.

■ If a known external frequency, f, was used to produce a particular standing wave, then the wave speed equation, $c = f\lambda$, can be used to calculate the speed of the wave, c.

■ For example, consider the third harmonic in Figure 4.58. If the length of the pipe was 53 cm, the wavelength was 71 cm. If the frequency which produced this standing wave was 480 Hz, then the wave speed was 340 m s^{-1}.

- Standing waves are commonly used in experiments to determine wave speeds.
- The speed of a transverse wave on a stretched spring depends on the applied tension and the mass per unit length. In stringed musical instruments either of these may be changed to produce notes of different frequency.
- In musical instruments using columns of air the speed of the wave cannot be changed, so different notes are achieved by changing the effective length of the air columns, or encouraging different harmonics.

QUESTIONS TO CHECK UNDERSTANDING

60 A sound standing wave is created in a 95.0 cm pipe which is open at both ends.

   a   Assuming the speed of sound was 336 m s⁻¹, calculate the frequency of the first harmonic.

   b   Another standing wave was detected at a frequency of 710 Hz. Which harmonic was this?

61 When a guitar string is plucked the whole string oscillates and the dominant frequency is the first harmonic.

   a   Explain what must happen to the string to make a louder sound of the same frequency.

   b   Explain why a different string will be needed to produce a lower pitched note.

62 Figure 4.60 shows an experiment designed to investigate how the speed of a wave along a stretched string varies with tension.

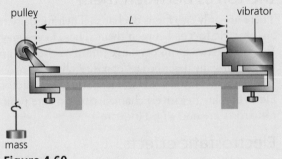

   a   Which harmonic is shown in the figure?

   b   If $L = 66$ cm, what is the wavelength shown?

   c   If the frequency of the vibrator was 25 Hz, what was the speed of the wave?

   d   What frequency would be needed to produce the fifth harmonic under these conditions?

   e   How could the tension be increased, and would the wave speed increase or decrease?

Figure 4.60

Expert tip

After the wave nature of electrons had been discovered, it was realized that an electron could be visualized as a standing wave within an atom. This was a major advance in atomic physics. This is discussed in Section 12.1 (HL).

NATURE OF SCIENCE

■ Common reasoning process

Historically, the study of standing waves has been closely linked to music. More than 2500 years ago, Pythagoras and others are credited with the foresight of believing in patterns and harmonies in the world around them and the objects in the night skies. Notably this included looking for patterns in the harmonies of musical notes. Such beliefs, original at the time, reflect the origins of our modern understanding of science: that there is a 'scientific method' which is common to all reasoning and experimentation in science. In particular, mathematics is often described as the 'language of science', but 2500 years ago the idea that there was a mathematical basis to the art of music was unacceptable to many at the time.

5.1 Electric fields

Essential idea: When charges move an electric current is created.

Charge

■ Electric **charge** is a fundamental property of some sub-atomic particles. Particles or objects with an overall charge are described as being 'charged'.

■ Identifying two forms of charge and the direction of the forces between them

■ There are only two forms of charge. Their fundamental property is that there are **electric forces** between them. These forces may be attractive or repulsive. We call the two forms of charge **positive charge** and **negative charge** and they are usually described as 'opposite' charges.

■ The most well-known sub-atomic particles are protons, neutrons and electrons. **Electrons** are negatively charged, **protons** are positively charged and **neutrons** are uncharged (**neutral**).

■ Electrostatic effects

■ It should be clear that all macroscopic objects contain countless numbers of charged particles. Usually the number of positive charges and the number of negative charges are about equal, so that overall we may describe the objects as being neutral.

■ It is usually not difficult to transfer some 'free' electrons from one object to another (by friction, for example), so that the object with extra electrons becomes negatively charged overall, while the other object (which has lost some electrons) then has an overall positive charge. (The total number of charges in the system remains constant: see *conservation of charge* below.)

■ Most everyday observations of electrostatic effects can be explained in terms of the transfer of electrons during the process of **charging**.

■ There are forces between charged objects, but it is also common for there to be forces between charged objects and *uncharged* objects. This is because some electrons in the uncharged object may be attracted to, or from, the area nearest the charged object.

■ Charged objects do not usually remain charged. This is because electrons will readily flow to, or from them to the Earth (or other objects). If this is done deliberately we talk about **earthing** an object. When an object loses its charge, we refer to it as being **discharged**.

■ Measurement of charge

■ Charge is given the symbol q (although some sources use Q, which may be confused with thermal energy).

■ The unit for charge is the **coulomb**, C. This is a relatively large unit, so that *microcoulombs*, μC (10^{-6} C), and *nanocoulombs*, nC (10^{-9} C), are also in common use.

■ One coulomb is the charge that flows past a point in one second if the current is one amp ($\Delta q = I\Delta t$; see later)

■ The smallest possible amount of charge on a free particle is equal to the charge on one electron, $e = -1.6 \times 10^{-19}$ C. A proton has charge of the same magnitude, but it is positively charged.

■ All (free) charge comes in multiples of this basic quantity. For this reason, charge is said to be *quantized*.

■ The conservation of charge means that if, for example, the charge on an object decreases, something else must have gained an equal charge.

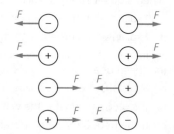

Figure 5.1

> **Key concept**
>
> Electric forces exist between all charged particles or objects. Opposite charges attract, like charges repel. See Figure 5.1.

> **Key concepts**
>
> The total charge in any isolated system is constant (the **law of conservation of charge**).

> **Expert tips**
>
> It might seem sensible to make the charge on the electron the basic unit for charge (rather than the coulomb), but that would be inconveniently small and inappropriate for macroscopic measurements.
>
> Quarks (Chapter 7) have charges of $\pm\frac{1}{3}e$ or $\pm\frac{2}{3}e$, but they do not exist as free, individual particles/charges.

QUESTIONS TO CHECK UNDERSTANDING

1 Draw a model of a neutral hydrogen atom with one proton and one electron. Label the particles with the signs and magnitudes of their charges and add vector arrows to show the relative sizes and directions of electric forces.

2 A lithium atom contains three protons and three electrons.

   a What is its total positive charge?

   b What is its total negative charge?

   c If it becomes ionized by the removal of one electron, what is its overall charge?

3 Explain how it is possible for dry hair to become positively charged by brushing.

4 Explain with the help of a diagram, how it is possible for a charged plastic rod to attract a small uncharged piece of paper.

Electric field

Revised ☐

- All electric charges create electric fields around themselves (where other charges will experience forces).

- The definition of electric field strength refers to a 'small test charge' because it must not have any properties which would upset the field that it is trying to measure.

- It may help understanding of electric fields (in which charges experience forces) to compare them with gravitational fields (in which masses experience forces). Electric field strength, E, is a concept analogous to gravitational field strength, g.

- However, gravitational fields always produce attractive forces, so that the direction of the field is obvious, whereas electric forces may be attractive or repulsive. For this reason, the definition of electric field strength has to refer to the *sign* of a charge. The direction of an electric field is *chosen* to be the direction of the force on a *positive* charge.

- Electric field strength is a vector quantity. If two or more fields exist at the same location, the resultant field strength can be determined by vector addition. Figure 5.2 shows the graphical method for determining the resultant of two separate fields, E_1 and E_2, which are not aligned.

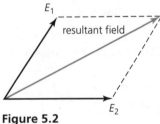

Figure 5.2

> ### Key concepts
>
> An **electric field** is a region of space in which any charge would experience an electric force.
>
> The **electric field strength, E,** at a point is defined as the force per unit charge (1 C) that would be experienced by a small *positive* **test charge** placed at that point. $E = \frac{F}{q}$; its units are $N\,C^{-1}$
>
> (or $V\,m^{-1}$).

> ### Expert tip
>
> Physicists use the term *field* to describe spaces in which forces can be experienced without any physical contact or the need for any intermediate medium. Field is a difficult concept to understand and sometimes the phrase '*forces acting at a distance*' is used. The most commonly discussed fields are gravitational, electric and magnetic.

■ Electric field diagrams

- We can represent electric fields by drawing field lines.
- The most basic electric fields are the **radial fields** around isolated point charges. Figure 5.3 shows the two possibilities

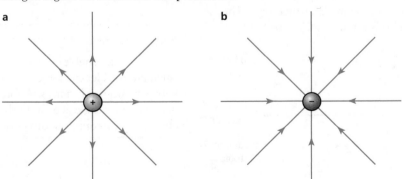

Figure 5.3

> ### Key concept
>
> **Electric field lines** show the direction of the force that would be exerted on a positive test charge if it was placed at that point (from positive to negative). A field is strongest where the lines are closest together.

- There are four other very common types of field. These are shown in Figure 5.4. Figures 5.4a and 5.4b show the fields around pairs of charges. The arrangement in Figure 5.4a shows opposite charges and is called a *dipole*. The charged sphere shown in Figure 5.4c behaves as if all of its charge was at its centre. Finally, Figure 5.4d shows the **uniform field** created between parallel charged metallic plates.

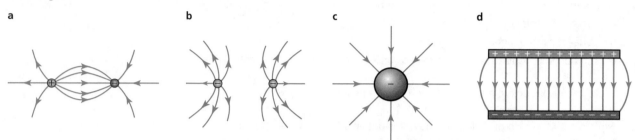

Figure 5.4

Solving problems involving electric fields

QUESTIONS TO CHECK UNDERSTANDING

5 Consider the diagrams in Figure 5.4. Which of the diagrams includes a place where the electric field is zero (a 'neutral point')?

6 a What is the magnitude of the electric field strength at a place where a charge of $+4.7 \times 10^{-8}$ C experiences a force of 9.3×10^{-5} N?

   b If the charge was replaced with another charge of -1.9×10^{-8} C in the same field, how would the force change?

7 Figure 5.5 shows a positively charged metal sphere suspended over a negatively charged metal plate. Use a copy of the diagram to represent the electric field produced by this arrangement. (Field lines are always perpendicular to conducting surfaces.)

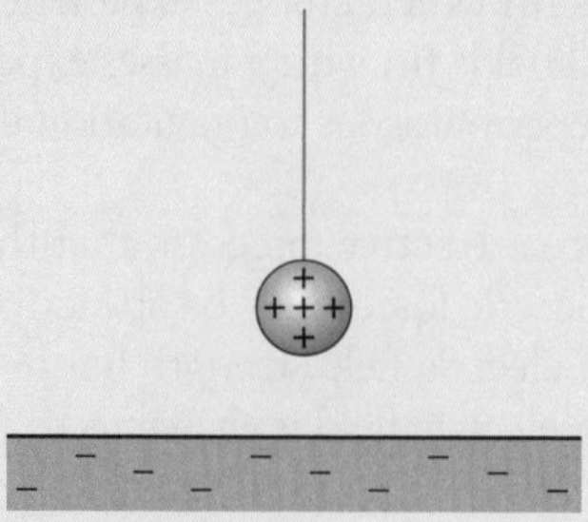

Figure 5.5

8 What is the magnitude of the resultant electric field when fields of $+3.4 \times 10^4$ N C^{-1} and -7.1×10^4 N C^{-1} combine:

   a if the fields are aligned

   b if the fields are perpendicular to each other?

Coulomb's law

Revised ☐

- The magnitude of the force, F, between two isolated point charges, q_1 and q_2, is proportional to the product of the charges and inversely proportional to the square of the distance, r, between them. See Figure 5.6, which represents repulsive forces between similar charges. This is known as **Coulomb's law:**

> **Key concept**
>
> *Coulomb's law* enables us to calculate the electric force between two *point charges*. A force of equal magnitude acts on both charges, but in opposite directions.

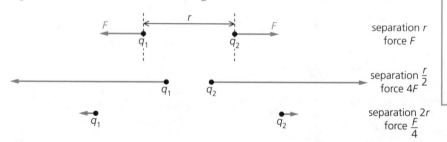

separation r
force F

separation $\frac{r}{2}$
force $4F$

separation $2r$
force $\frac{F}{4}$

Figure 5.6

- Coulomb's law: $F = k\dfrac{q_1 q_2}{r^2}$, where k is known as the **Coulomb constant**. Its value is 8.99×10^9 N m^2 C^{-2}.

- If both charges have the same sign the force is positive (repulsive). If the charges have opposite signs the force is negative (attractive).
- Coulomb's law can also be used with charged spheres, which are surrounded by electric fields that behave as if all the charge was concentrated at the centre of the spheres (Figure 5.4c). r is then the distance between the centres of the two spheres.
- Coulomb's law as expressed above assumes that the space between and around the two charges is a vacuum (or air). The Coulomb constant can be written in the expanded form $k = \dfrac{1}{4\pi e_0}$ to make this origin clearer. ε_0 is a constant which describes the electric properties of vacuum. It is called the **permittivity of free space** (vacuum). Its value is $8.85 \times 10^{-12}\,C^2\,N^{-1}\,m^{-2}$. In Coulomb's law, the value of ε_0 controls the magnitudes of the forces between charges in vacuum.
- Coulomb's law can therefore be alternatively written as: $F = \dfrac{1}{4\pi e_0}\dfrac{q_1 q_2}{r^2}$.
- If the charges were surrounded by any other medium (water for example), the forces between them would be reduced and this is represented in the equation above by changing ε_0 to ε, the *permittivity of the medium*.
- Every medium has a permittivity of greater value than ε_0. The permittivity of air may be assumed to be the same as for vacuum.

> **Expert tip**
>
> Coulomb's law is mathematically very similar (analogous) to Newton's law of gravitation (Section 6.2).

> **Key concept**
>
> The electric properties of a particular medium are represented by its **permittivity**, ε.

> **Expert tips**
>
> We can determine the radial electric field strength around a point charge (q_1) by using Coulomb's law. The field, E, at a distance r, equals F/q_2, where q_2 is a test charge placed at that point. But from Coulomb's law $F = kq_1q_2/r^2$, so that $E = kq_1/r^2$. There is more about this in Chapter 10 (HL).
>
> The ratio $\varepsilon/\varepsilon_0$ is sometimes called the *relative permittivity* of a medium (or its dielectric constant: see Section 11.3, for HL students).
>
> *Permeability* is a magnetic property of a medium comparable to its electric *permittivity* (see Section 5.4). It is a measure of their importance that these two constants control the value of the speed of electromagnetic waves.

■ Solving problems involving Coulomb's law

QUESTIONS TO CHECK UNDERSTANDING

9 What is the force between a proton and an electron in a hydrogen atom (assume separation $= 5.3 \times 10^{-11}\,m$)?

10 There was a force of $6.3 \times 10^{-8}\,N$ between two charged spheres, each of radius 2.8 cm when the separation of their surfaces was 10.0 cm in air. If one sphere had a charge of 8.7 nC, what was the charge on the other?

11 a If the measurements described in the previous question were repeated with the spheres placed in a container of carbon dioxide (rather than air), would the force remain the same, increase or decrease?

  b Explain your answer with a calculation. The electric permittivity of carbon dioxide is $1.4 \times 10^{-11}\,C^2\,N^{-1}\,m^{-2}$.

12 Determine the electric field strength at a distance of 25 cm from a point charge of 10 nC.

Electric current

`Revised` ☐

- An **electric current** is a flow of charged particles. In electrical circuits the moving charges are usually electrons.

■ Direct current (dc)

- Currents which always flow in the same direction around a circuit are called **direct currents (dc)**. The direction is shown with an arrow that *always* points from the positive terminal (of the energy source) around the circuit to the negative terminal. This is called the direction of **conventional current**.

> **Expert tip**
>
> Currents in liquids and gases usually involve the movement of *ions*, but that is *not* a part of this course. An ion is an atom or molecule that has an overall charge because it lost or gained one or more electrons.

(Although electrons in circuits actually flow the other way.) Even if the magnitude of a current changes, it is still described as a direct current if the flow is always in the same direction. See Figure 5.7.

■ **Alternating currents (ac)** repeatedly change direction. Most commonly, alternating currents keep a constant frequency, for example the *mains* ac frequency in many countries is 50 Hz. This chapter concentrates on dc, but there is more about ac in Chapter 11 (HL). (The term **mains electricity** is used for electrical energy supplied to many different homes by cables from power stations.)

■ Current is given the symbol *I*. Current = charge/time. $I = \dfrac{\Delta q}{\Delta t}$.

■ The unit of current is the **ampere (amp)**, A. $1\,\text{A} = 1\,\text{Cs}^{-1}$.

■ Since the charge on the electron is $-1.6 \times 10^{-19}\,\text{C}$, a flow of 1 C of charge past a point in a metallic conductor in a circuit every second involves the movement of 6.25×10^{18} electrons.

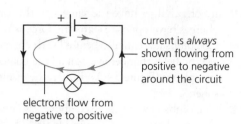

current is *always* shown flowing from positive to negative around the circuit

electrons flow from negative to positive

Figure 5.7

Identifying sign and nature of charge carriers in a metal

■ A material through which charges can flow is called a **conductor**. If charge cannot flow through a material it is called an **insulator**. No material is a *perfect* conductor or insulator (with the exception of very cold *superconductors*).

■ The difference between solid conductors and insulators is explained in terms of the number of **free electrons** (sometimes called *delocalized electrons*) that are able to move through the material. In good conductors, like metals, some of the outer electrons (of the atoms) are not permanently attached to particular atoms, but they can move around freely, like the random motion of molecules in a gas.

■ This dynamic situation is not easy to represent in a single diagram, but Figure 5.8 shows an idealized representation of a current carried through a metal by free electrons. The actual motions of the electrons are much more erratic than the *net* movements shown in the figure. The electrons in a current have to move past the vibrating ions of the metallic structure and this explains why the conductor has electrical *resistance* to the flow of current (see Section 5.2).

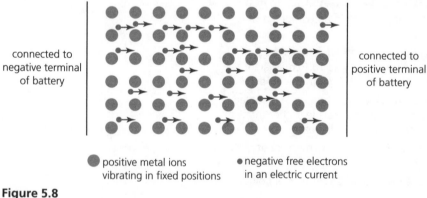

connected to negative terminal of battery

connected to positive terminal of battery

● positive metal ions vibrating in fixed positions   ● negative free electrons in an electric current

Figure 5.8

■ Electric currents are measured by **ammeters** placed *in series* in the circuit. See Figure 5.9, in which a cell, a lamp and an ammeter are connected in a simple series circuit.

Identifying drift speed of charge carriers

■ The **drift speed**, *v*, of electrons (or other charge carriers) in an electric current is their *net* speed along the conductor. It should not be confused with their instantaneous very high speeds in random directions (motions which they have even when there is no current flowing).

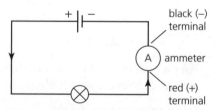

black (–) terminal

A ammeter

red (+) terminal

Figure 5.9

■ The magnitude of a current in a conductor depends on the drift speed, v, of the charges, but also on the magnitude of the charges, q, the **charge density**, n (number of charges in a cubic metre of the conductor) and the cross-sectional area of the conductor, A. Figure 5.10 shows a section of a cylindrical shaped conductor (for example, a metal wire).

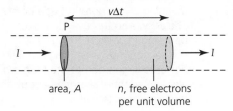

Figure 5.10

■ During a time interval Δt an electron of charge q will move a net distance of $v\Delta t$. So that the number of electrons flowing past a point P in time Δt will be $nAv\Delta t$ (charge density × volume).

■ Using $I = \dfrac{\Delta q}{\Delta t}$ leads to the **drift speed equation**: $I = nAvq$

■ Solving problems involving current and charge, and using the drift speed equation

■ The charge density for metals is very high. For example, the number of copper atoms in a cubic metre is 9×10^{28} and if, on average, each copper atom supplies approximately one free electron, the charge density for copper is also about 9×10^{28} electrons per cubic metre.

■ When typical values for I, A and n are inserted into this equation, we discover that the drift speed for electrons is surprisingly low.

■ **Semiconductors** like silicon contain *much* fewer charge carriers. This means that if the same current passes through a semiconductor and a metal in the same series circuit, the charges must move *much* faster in the semiconductor.

QUESTIONS TO CHECK UNDERSTANDING

13 a How much charge flows through a light bulb in one hour if the current is 0.24 A?

b How many electrons pass through the bulb every second?

14 Calculate the drift speed of electrons in a copper wire of diameter 1 mm when there is a direct current of 1.0 A (assume the charge density for copper = 8.5×10^{28} electrons m⁻³).

15 Suggest how the drift *velocity* of electrons may change when a current is alternating.

Energy and charge

Revised ☐

■ Electrical energy is transferred in circuits by moving charges, and that energy is then transferred into other forms by the *components* in a circuit (for example, lamps, motors and heaters).

■ We now need to introduce the very important concept of energy transferred to, or from, charges:

■ Potential difference

■ P.d is sometimes called **voltage** because the unit of p.d. is the **volt**, V.

■ A p.d. of one volt means that one joule of energy is transferred by one coulomb of charge moving between the two points ($1\,V = 1\,J\,C^{-1}$).

■ A p.d. *supplied* by a battery, or other power source, is needed across any circuit in order to make a current flow through it. The sum of the p.d.s 'used' by the components around a series circuit must add up to the p.d. supplied. See next point.

Key concepts

Metals contains many negatively charged *free electrons* which are not tied to specific atoms and which move around randomly. An electric current exists when they are made to also move (drift) along a wire in the same direction as each other.

The value of the current can be related to the electron *drift speed* and the charge density.

$I = nAvq$

Common mistakes

The drift speed of electrons in currents through metal wires may seem low when we consider how quickly most electrical devices start working after we turn them on. To understand why, we need to consider that currents do not start at the power supply or the switch, rather we should imagine that all electrons in the circuit start moving at that same moment (although this cannot be perfectly true).

Key concept

The energy that would be transferred if one unit of charge (1 C) moved between two points is called the **potential difference (p.d.)**, V, between those points.

p.d. = energy transferred (work done)/ charge: $V = \dfrac{W}{q}$.

■ The p.d. between any two points in a circuit can be measured with a **voltmeter** connected between the points (*in parallel across* the component being checked). In Figure 5.11, $V_s = V_1 + V_2$

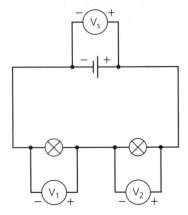

Figure 5.11

Expert tip

The term *potential difference* between two points may cause some confusion in this section, especially if it raises the question: what is the meaning of the *potential* at a point? The answer to this question is to be found in Section 10.1 (HL). In this section, however, it is probably better to simply accept the term p.d. (voltage) without worrying about the meaning of potential.

■ The concept of potential difference is not limited to electrical circuits. It can be applied to any electric field. For example in a storm there could be a potential difference of 10^8 V between a cloud and the ground.

■ Consider the electric field, E, shown in Figure 5.4d, in which a charge q would experience a force F. If the plates are separated by a distance d and have a p.d. of V across them, then the work done in moving the charge from one plate to the other $= Fd = Vq$. Rearranging gives $\dfrac{F}{q} = E = \dfrac{V}{d}$. This gives us a convenient way of calculating the uniform electric field between parallel plates. The units $V\,m^{-1}$ are equivalent to $N\,C^{-1}$.

A unit for energy in atomic physics: the electronvolt

■ When a charged particle is accelerated from rest by a p.d., the work done ($W = Vq$) can be equated to the kinetic energy ($\frac{1}{2}mv^2$) in order to determine the particle's speed.

■ If a free electron was accelerated through a p.d. of 1000 V, the (kinetic) energy that it would gain = p.d. × charge = 1000 V × (1.6 × 10^{-19}) C = 1.6 × 10^{-16} J. This is such a common type of calculation in atomic physics that the energy transfer when an electron is accelerated by 1000 V is simply said to be 1000 *electronvolts* (avoiding the need for any calculation, or the use of very small numbers).

■ The conversion between electronvolts and joules: $W(eV) = \dfrac{W(J)}{e}$.

■ The unit electronvolt is used widely throughout atomic physics (in preference to joules). It can be used with *any* particle or radiation (not just accelerated electrons). The units keV and MeV are also in very common use.

Key concept

The **electronvolt**, eV, is defined as the unit of energy which is equal to that which is transferred when a charge of 1.6 × 10^{-19} C moves through a p.d. of 1 V.

Solving problems involving potential difference and charge/Calculating work done in an electric field in both joules and electronvolts

QUESTIONS TO CHECK UNDERSTANDING

16 a How much work is done in an electric field when a charge of +5.6 nC moves though a potential difference of: **i** +250 V, **ii** −250 V?

   b In which case is the work done *on* the charge?

17 A p.d. of 5000 V was connected across parallel metal plates which were separated by a distance of 8.3 cm.

   a The top plate was negatively charged. Sketch the arrangement and show the electric field lines.

   b What was the magnitude of the uniform electric field created between the plates?

c Explain how it might be possible for a tiny charged oil drop to remain stationary between the plates.

d What would be the magnitude of the electric force if the drop had a charge of 1.12×10^{-18} C?

18 What is the potential difference across a battery which transfers 135 J of energy in 5 s to a current of 3 A?

19 A lamp in a circuit has a potential difference of 4.5 V across it. If the lamp has a power of 2.5 W, how many coulombs of charge flow through it every second?

20 a What p.d. is needed to accelerate an electron from rest to a kinetic energy of 500 eV?

b Determine the final speed of the accelerated electron.

21 A doubly charged positive ion is accelerated by a p.d. of −5.6 kV.

a What is its final kinetic energy: **i** in eV, **ii** in J?

b What assumption did you make?

22 An alpha particle emitted by an unstable nucleus (see Chapter 7) has a kinetic energy of 2.8×10^{-13} J. Express this energy in MeV.

NATURE OF SCIENCE

■ Modelling

Understanding physics requires that students accept *models* of reality. Models may be physical representations, drawings, mathematical equations, theories, etc. All models aim to simplify, to help understanding and to make predictions. In nearly every topic in physics, students need to be able to explain macroscopic observations by using microscopic models about the behaviour of atoms, ions and molecules. The model of electrical conduction of free electrons in solids is a good example from this chapter. Since particle behaviour cannot be directly observed, in order to understand electricity, students need to be able to use models of the microscopic world that they learn from their teachers or other resources, with no clues from their own senses. But these models have stood the test of time.

5.2 Heating effect of currents

Revised ☐

Essential idea: One of the earliest uses for electricity was to produce light and heat. This technology continues to have a major impact on the lives of people around the world.

Heating effect of current and its consequences

Revised ☐

■ Whenever any electric current passes through a conductor, some energy will be transferred to internal energy which then spreads away from the conductor as thermal energy. All currents have 'heating effects'.

■ The amount of thermal energy dissipated depends on the magnitude of the current and the *resistance* of the conductor.

■ In many circuits we may wish to keep the energy that is dissipated as low as possible, but on other occasions we wish to use electricity to make things hotter, or to produce light.

Top 2
Y.4

Circuit diagrams

- **Circuit diagrams** show how power sources are connected with metallic wires (*connecting leads*) to devices which transfer energy usefully. There are two basic kinds of circuit: series and parallel.
- Placing circuit components in **series** means that they are connected one after another, with no alternative paths for an electrical current. The same current flows through all the components.
- If components are connected in **parallel** it means that the current divides because it can take two or more different paths between the same two points in the circuit.
- Some circuits contain both series and parallel sections.

▨ Drawing and interpreting circuit diagrams

- This is an important skill required throughout this chapter and elsewhere in the course.
- Students need to be familiar with all the standard circuit symbols shown in the IB *Physics data booklet*. See Figure 5.12. If a symbol is forgotten it may be acceptable to use a rectangle with its name written inside.

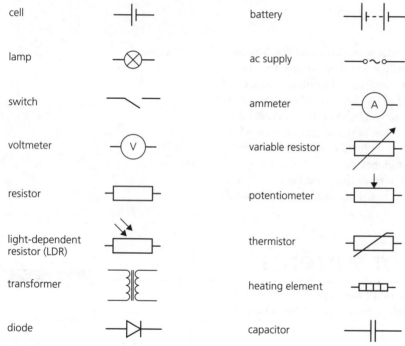

cell	⊣⊢	battery	⊣⊢--⊣⊢
lamp	—⊗—	ac supply	—o∿o—
switch	—╱—	ammeter	—(A)—
voltmeter	—(V)—	variable resistor	variable resistor symbol
resistor	—▭—	potentiometer	potentiometer symbol
light-dependent resistor (LDR)	LDR symbol	thermistor	thermistor symbol
transformer	transformer symbol	heating element	heating element symbol
diode	—▷|—	capacitor	—⊣⊢—

Figure 5.12

Kirchhoff's circuit laws: first law

- These two laws are mathematical summaries of conservation principles applied to circuits.
- Because of the *law of conservation of charge*, the total current coming into any junction in an electric circuit must equal the total current leaving the same junction. $\Sigma I = 0$ **(junction)**. This is *Kirchhoff's first law*. Currents arriving at a junction may be considered as positive and currents leaving as negative.
- Figure 5.13 shows an example.
- Kirchhoff's second law ($\Sigma V = 0$ (loop)) is covered later in this section.

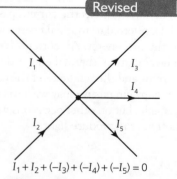

$$I_1 + I_2 + (-I_3) + (-I_4) + (-I_5) = 0$$

Figure 5.13

Key concept

Kirchhoff's first law: the total current arriving at any junction is equal to the total current leaving the same point.

Resistance expressed as $R = \dfrac{V}{I}$

- In general terms, electrical resistance is a measure of how difficult it is to pass an electric current through something.

- The resistance of a component may change for a number of different reasons. In particular, significant temperature changes usually affect the resistance of a component.

- A component which is designed to have a certain resistance is called a **resistor**. *Variable resistors* are widely used for varying currents and voltages in circuits.

> **Key concept**
>
> **Resistance**, R, is the ratio of the p.d. across a component to the current through it: $R = \dfrac{V}{I}$. The unit of resistance is the **ohm**, Ω. $1\,\Omega = 1\,V\,A^{-1}$.

■ Solving problems involving, current, Kirchhoff's first circuit law, potential difference and resistance

QUESTIONS TO CHECK UNDERSTANDING

23 Consider Figure 5.13. If I_1, I_2, I_3 and I_4 were all 1 A, what was I_5?

24 Are the appliances in your home connected in series or parallel? Justify your answer.

25 a What is the resistance of a 230 V domestic light bulb if the current through it is 0.11 A?

　 b Suggest why the current through the bulb may have been higher at the moment that it was first turned on.

　 c What current would flow through the bulb if it was connected to 110 V?

　 d What assumption did you make in answering part **c**?

26 What p.d. will make a current of 2.4 mA pass through a 4.7 kΩ resistor?

$\dfrac{V}{I}$ characteristics

- A knowledge of how different p.d.s affect current is important information about any component. The simplest way of investigating this is shown in Figure 5.14. The p.d. can be changed by altering the output of the low voltage dc supply (or, if batteries are used, by varying the number of cells). Other methods are discussed later.

- The results of such experiments are usually shown graphically, and are commonly known as $\dfrac{V}{I}$ **characteristics**.

- Figure 5.15 shows the simplest example, where the current is proportional to the p.d. This is an idealized result, but it may be assumed to be accurate for metals at constant temperature. It is a representation of *Ohm's law*:

■ Ohm's law

- A component for which Ohm's law applies is said to be **ohmic**. An ohmic component has a constant resistance.

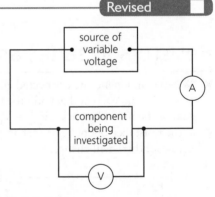

Figure 5.14

> **Key concept**
>
> **Ohm's law** states that the current through a metallic conductor is proportional to the potential difference across it, if the temperature remains constant.

■ Identifying ohmic and non-ohmic conductors through a consideration of the $\dfrac{V}{I}$ characteristic graph

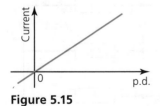

Figure 5.15

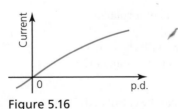

Figure 5.16

> **Common mistake**
>
> Resistance is not equal to the gradient of a current—p.d. graph (unless the component is ohmic).

- A component to which Ohm's law does *not* apply is called *non-ohmic* and it does not have a constant resistance. Figure 5.16 shows a typical example, which could, for example, represent the behaviour of a **filament lamp**. A filament emits more light as the current increases and it gets hotter, but this also increases its resistance.

QUESTIONS TO CHECK UNDERSTANDING

27 **a** Sketch a $\frac{V}{I}$ characteristic (for voltages up to 10V) for a 5 Ω ohmic resistor.

  b Add to your sketch a line which could represent a filament lamp which had a resistance of 5 Ω when the voltage was 6 V.

28 Figure 5.17 represents the $\frac{V}{I}$ characteristic of a component called a thermistor, with increased temperature as the current rose.

  a Describe the behaviour shown in the graph.

  b Calculate the resistance of the component when the p.d. was: **i** +2.0V, **ii** −4.0V.

  c Sketch the $\frac{V}{I}$ characteristic of a diode which only allowed current to pass through it in one direction.

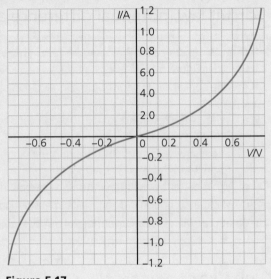

Figure 5.17

Resistors in series and parallel

■ When components are connected *in series* (see Figure 5.18) the same current flows through each of them and the total p.d. across all of them is equal to the sum of the individual p.d.s: $V_{total} = V_1 + V_2 + V_3$.

■ If the resistance of one resistor changes, the current will change and the p.d.s across all three will change.

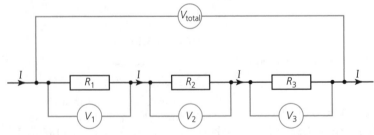

Figure 5.18

■ The total resistance of two or more *resistors in series* can be calculated from $R_{total} = R_1 + R_2 + ...$

■ When components are connected *in parallel* (see Figure 5.19) they each have the same p.d. across them and the total current is equal to the sum of the individual currents $I_{total} = I_1 + I_2 + I_3$

■ The total resistance of two or more *resistors in parallel* can be calculated from $\frac{1}{R_{total}} = \frac{1}{R_1} + \frac{1}{R_2} + ...$

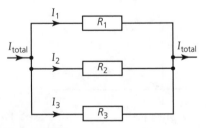

Figure 5.19

■ Describing ideal and non-ideal ammeters and voltmeters

- ■ Look again at Figures 5.9 and 5.11, which show how ammeters and voltmeters should be connected in a circuit. *Ideal* meters, of course, will not affect the quantities that they are measuring.
- ■ Practical (non-ideal) ammeters usually have very low resistances (maybe $10^{-3}\,\Omega$) and digital voltmeters have very high resistances (maybe $10\,\text{M}\Omega$ or more). Under most circumstances such meters may be considered to be 'ideal', because their resistances are very different from the resistances of other components in a circuit.
- ■ However, occasionally, in high resistance or very low resistance circuits, the resistances of the meters will have a significant effect on the circuit, so that their resistances affect the readings that they display.

> ### Key concept
> An **ideal ammeter** has zero resistance, an **ideal voltmeter** has infinite resistance.

■ Solving problems involving potential difference, current and resistance

QUESTIONS TO CHECK UNDERSTANDING

29 Three resistances have values of $100\,\Omega$, $200\,\Omega$ and $300\,\Omega$.

  a Draw the combination of these three resistors which has the greatest resistance. What is their combined resistance?

  b Draw the combination of these three resistors which has the least resistance. What is their combined resistance?

30 Show how three equal resistances, each $2\,\Omega$, can be connected together to produce an overall resistance of $3\,\Omega$.

31 Figure 5.20 shows five identical lamps in a circuit. Assume that their resistances are equal and constant.

  a If each lamp has a resistance of $6.0\,\Omega$, determine the overall resistance of the circuit.

  b If the battery supplies a p.d. of $12.0\,\text{V}$, calculate the currents through:
   i lamp A, **ii** lamp C, **iii** lamp E.

  c If lamp D stops working, how will that affect the brightness of the other lamps?

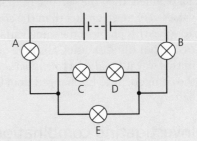

Figure 5.20

32 Figure 5.21 shows a voltmeter connected with a switch across one of two one million ohm resistors connected in series to a $6.0\,\text{V}$ cell.

  a If the voltmeter also has a resistance of one million ohms, what voltage reading will it display when the switch is closed?

  b What would the voltages across the two resistances be when the switch was open?

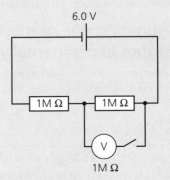

Figure 5.21

33 Figure 5.22 shows three resistors connected to a battery supplying a p.d. of V_s.

  a What value of the variable resistance will produce a reading of $\frac{V_s}{2}$ on the voltmeter?

  b What are the maximum and minimum readings on the voltmeter when the variable resistor is adjusted?

34 Consider Figure 5.14.

  a If the voltmeter reading was $4.7\,\text{V}$ and the ammeter read $1.3\,\text{A}$, what was the value of the resistance?

  b If a student incorrectly connected the meters, with the ammeter where the voltmeter is shown, and the voltmeter where the ammeter is shown, what readings would be seen on the meters? (Assume that they were ideal meters.)

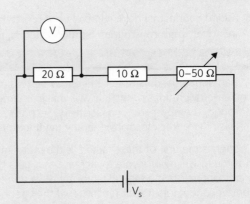

Figure 5.22

Experiments to investigate resistance

Revised

- The circuit shown in Figure 5.14 can be used to determine resistance values and how they may change. But, there is an alternative method which uses a *fixed* voltage supply and a *variable resistor* to provide varying voltages. See Figure 5.23.

- The p.d. from the battery is shared by the component under investigation and the variable resistor. If the value of the variable resistance is (for example) increased, the current in the series circuit will decrease and the p.d. across the component will also fall.

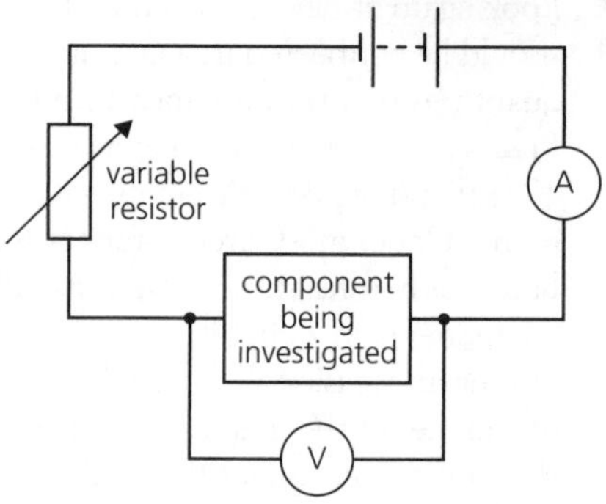

Describing practical advantages of a potential divider over a series resistor in controlling a simple circuit

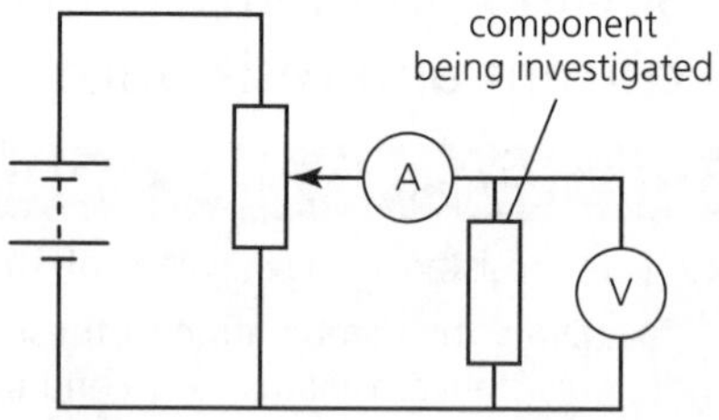

- In Figure 5.23, in order for the component to be able to get the full range of p.d.s available from the battery, the resistance of the variable resistor would need to change from zero to infinite. More realistically, if low values of p.d. are required, then the variable resistance would need to be able to increase to values very much greater than the component.
- To solve this problem the same variable resistance can be connected as a **potential divider**, using all three of its terminals. See Figure 5.24. Used in this way it is called a *potentiometer*. This circuit enables the p.d. across the component to be changed from 0 V to the maximum available from the battery.

Investigating combinations of resistors in parallel and series circuits

- It is possible to experimentally confirm the equations for series and parallel combinations of resistors by using the circuits described above.

Investigating one or more of the factors that affect resistance experimentally

- The same circuits can be used to investigate how the resistance of a wire depends on its length, cross-sectional area or temperature.

Figure 5.23

Figure 5.24

> **Key concept**
>
> Connecting a three terminal variable resistor across a battery makes a potential divider capable of fully varying the voltage input to the rest of the circuit.

Resistivity

Revised

- The resistance, R, of a length of metal wire is proportional to its length, L, and inversely proportional to its cross-sectional area, A (at constant temperature).

- In general, for a regularly shaped specimen of the same material, $R = \dfrac{\rho L}{A}$.

- Good conductors have relatively low resistivities. Good insulators have relatively high resistivities. Semiconductors, like silicon, are in between.

> **Key concepts**
>
> The **resistivity**, ρ, of a material equals the resistance of a one metre length of cross-sectional area 1m²: $\rho = \dfrac{RA}{L}$
>
> The unit of resistivity is $\Omega\,m$.

> **Common mistake**
>
> The unit of resistivity is $\Omega\,m$, *not* $\Omega\,m^{-1}$.

Expert tip

When a material gets hotter there may be two effects on its resistivity. Firstly, increased vibrations of the ions will make it harder for the electrons to pass through, thereby increasing resistivity. Secondly, the extra energy supplied may release more free electrons, thereby tending to reduce resistivity.

In metals the first of these factors is the more important, so there is a tendency for the resistivity of a wire to increase as it gets hotter. But for insulators and semiconductors the second factor tends to dominate and they may be *much* better conductors (lower resistivity) at they get hotter.

If some metals are cooled to very low temperatures their resistivity falls to zero. They are then said to be **superconducting**.

Solving problems involving potential difference, current, resistance and resistivity

QUESTIONS TO CHECK UNDERSTANDING

35 Describe an experiment to investigate how the cross-sectional area of a wire affects its resistance.

36 Calculate the resistance of a 2.0 km length of aluminium cable of cross-sectional area 1.8 cm². The resistivity of aluminium is $2.8 \times 10^{-8}\,\Omega\,\text{m}$.

37 A student applied a p.d. of 1.46 V across the ends of a 98 cm length of an alloy of resistivity $1.1 \times 10^{-8}\,\Omega\,\text{m}$. If the diameter of the wire was 0.14 mm, what was the current in the wire?

38 Suggest how the charge density and resistivity of a material are related to each other.

39 Glass has a resistivity $\approx 10^{12}\,\Omega\,\text{m}$ at room temperature and is considered to be an insulator.

   a   What current does this predict for a glass rod of cross-sectional area 5 mm² and length 10 cm when connected to 12 V?

   b   A student tried to measure this current, but was unsuccessful. But when the rod was heated very strongly a current of 0.05 A was measured. Estimate the resistivity of the hot glass.

40 Consider Figure 5.23. If the component had a resistance of 10 Ω and the variable resistor was adjustable between 0 and 30 Ω, what range of p.d.s could be applied across the component using a 12 V battery?

41 Draw a diagram of a circuit with a potential divider which could be used to investigate the resistance of a 12 V, 0.2 A lamp. Label the components with suitable values.

Kirchhoff's circuit laws: second law

- Because of the law of conservation of energy, the sum of the p.d.s supplying energy to any circuit loop must equal the sum of the p.d.s across components used for transferring energy to other forms. $\Sigma V = 0$ **(loop)**. This is known as *Kirchhoff's second circuit law*.
- A circuit 'loop' means *any* path that starts and ends at the same point.
- Typically the p.d.s across the components will not be known directly, so *IR* must be used for each component, remembering the directions of the currents. (Unknown current directions can be guessed and the sign of the answers will indicate if the guesses were right or wrong.)
- If there is more than one source of power in a circuit loop (for example two separate batteries), one p.d. must be subtracted from the other if they are connected in 'opposition'.
- Consider Figure 5.25. Three circuit loops can be identified. In the lower loop the batteries are connected the same way around, so that the supply voltage is 17 V and V_2 must be 14 V. In the small loop the sum of the supply voltages is 15 V, which must equal V_1. In the outer loop the batteries are in opposition so that the total supply p.d. is 2 V, which equals $14 + 3 - 15$. (The 15 is negative because the current through it is being driven in the opposite direction.)

> **Key concept**
>
> **Kirchhoff's second law:** the sum of the p.d.s around any circuit loop always adds up to zero, but it is necessary to distinguish between components that are transferring energy to the current and those that are transferring energy from the current to other forms.

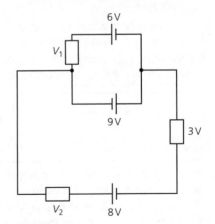

Figure 5.25

■ Solving problems involving Kirchhoff's circuit laws

■ Both laws can used together to determine currents flowing in circuits with more than one loop and more than one power supply.

QUESTION TO CHECK UNDERSTANDING

42 Determine all three currents in the circuit shown in Figure 5.26.

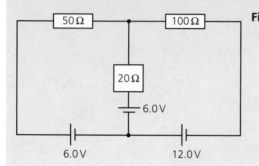

Figure 5.26

Power dissipation

■ The defining equations for p.d. and current can be combined to produce an equation for *electrical power* dissipation in a resistance, $P = VI$ (unit: watt, W).
■ The total energy supplied in a time t can be determined from **energy = VIt**.
■ Using $R = V/I$, it can be shown that the power is also given by: $P = I^2R = \dfrac{V^2}{R}$.
■ Energy delivered by electricity to our homes is bought and paid for in the unit of *kilowatt-hour* (rather than the much smaller unit of joules). One **kilowatt-hour**, kWh is the amount of energy transferred in a device of power one kilowatt in a time of one hour. 1 kilowatt-hour (kWh) = 3.60×10^6 J.

> **Key concept**
> Electrical power, $P = VI$
> (unit: watt, W)

■ Solving problems involving potential difference, current, power and resistance

QUESTIONS TO CHECK UNDERSTANDING

43 a What current flows through a 230 V domestic water heater which has a power of 2.5 kW?

   b How long will it take this heater to transfer 10^6 J of energy?

   c What is the resistance of the heater?

44 If an electrical power line had an effective resistance of 0.2 Ω km^{-1}, what power loss would occur in a 20 km cable if the current was 50 A?

45 a If the p.d. across a resistor is doubled, by what factor does the power dissipated change?

   b If a fixed p.d. is connected across a variable resistor, how does the power dissipated change if the resistance is doubled?

46 If the cost of electricity is 10 cents per kWh, how much will it cost every week to use a 150 W television for 4 hours a day?

Describing practical uses of potential divider circuits

■ **Sensors** are components that respond to a difference in a physical property with a corresponding change in resistance (or other electrical property). The resistance of a **light-dependent resistor (LDR)** decreases with light intensity; the resistance of most **thermistors** decreases as the temperature increases; the resistance of a *strain gauge* increases as it gets longer.
■ Sensors are often connected in series with another resistor and a power source (battery), so that they share (divide) the total p.d. When the resistance of the sensor changes, so too do the p.d.s, and the changing p.d.s can be used to turn

> **Key concept**
> The voltage across a sensor in a potential divider can be used to control the operation of another part of the circuit.

another part of the circuit on or off. Such arrangements are another example of *potential dividing* circuits.

■ Figure 5.27 shows an example. When the light intensity on the LDR increases, its resistance decreases and the p.d. across it falls, while the p.d. across the variable resistance rises by an equal amount. V_{out} can be used to control another part of the circuit (a light perhaps). The value of the variable resistor can also be changed in order to adjust V_{out}.

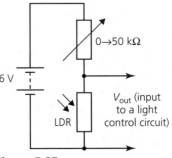

Figure 5.27

QUESTION TO CHECK UNDERSTANDING

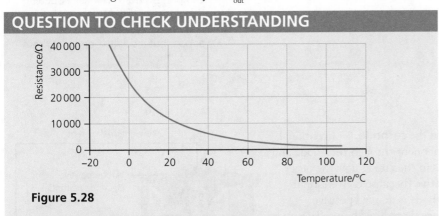

Figure 5.28

47 Figure 5.28 shows how the resistance of a certain kind of thermistor varies with temperature.

  a Describe this behaviour.

  b Estimate its resistance at 15 °C.

  c Draw a potential dividing circuit including a 12 V battery, a thermistor and a 0–50 kΩ variable resistor, that could be used as an input to a circuit designed to switch on a heater when the temperature went below a certain value.

  d If the variable resistor was set at a value of 25 kΩ, what was the voltage across it when the temperature was: **i** 15 °C, **ii** 100 °C?

  e Suggest why the value of the variable resistor might be changed.

NATURE OF SCIENCE

■ Peer review

In the modern scientific community the results of any new research are often published world-wide (perhaps after patenting!) and therefore subjected to very close and prompt scrutiny by the researchers' fellow scientists. The status and reputation of the researchers involved will not stop critical analysis. This is called *peer review*, and it is a key component of modern science. However, in the past things were very different. Apart from the fact that all communication was much slower, scientists tended to work more individually and their social status and reputation were considered important factors in judging their work. The initial rejection of Ohm's law (Barlow's ideas were given greater credence) is a good example of this prejudice.

5.3 Electric cells

Revised ▢

Essential idea: Electric cells allow us to store energy in a chemical form.

Cells

Revised ▢

■ **Electric cells** use chemical reactions to transfer energy to electric currents. A battery consists of two or more cells, although the term is also very commonly used for single cells.

■ The simplest cells have two electrodes, made from different conductors (usually metals or graphite), which are placed in an **electrolyte** (a solution or gel or paste that contains mobile ions). See Figure 5.29 for a basic example. An **electrode** is the name that we give to any conducting contact made with a non-metallic component (like an electrolyte) in a circuit. This physics course does not expect students to have knowledge of the chemical changes that occur inside a cell.

> **Key concept**
>
> Chemical reactions in an electric cell can transfer energy to an electric current flowing through it. In some cells, called **secondary cells**, the reactions can be reversed, so that the cell can be recharged and used again.

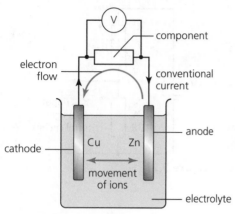

Figure 5.29

Common mistake

The direction of currents (or electrons) into or out of components with electrodes can cause confusion. This is usually because we need to distinguish between what is happening inside the component from what is happening in the rest of the circuit. Conventional current flows from a circuit *into* an anode, but inside the device it flows *out* of the anode.

- The electrodes are called the **anode** and the **cathode**. Conventional current flows from the circuit into a component through the anode and out from the cathode. The free electrons in the circuit flow the other way. On a cell the anode is marked negative (because conventional positive charge flows into it) and the cathode is marked positive.
- The useful voltage (*terminal potential difference*) produced by a cell depends on the materials used, but not on the physical dimensions of the cell.
- Figure 5.30 shows a typical 'dry cell' which might be used in a torch. (Details do not need to be remembered.)
- If a cell has to be discarded after the chemical reactions have finished, it is known as a **primary cell**. These are the cheapest types of cell, but the disposal of large quantities of batteries causes pollution problems.
- The useful energy stored in a cell or battery is not usually given in joules. Instead, battery manufacturers use the units of Ah or Wh (both assume the voltage is constant). For example, a 12V, 150Ah car battery should be able to provide a 1A current at 12V for 150h. Using energy = *VIt* shows that this is equal to 6.48×10^6J. The same energy could supply, for example, 3A for 50h. This amount of energy is equivalent to 1800Wh.

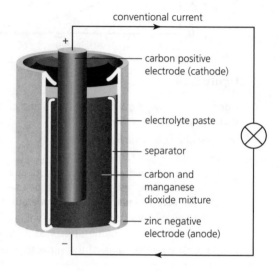

Figure 5.30

Secondary cells

Revised

- In some types of cell it is possible to reverse the chemical changes which transferred the energy and thereby enable the re-use of the cell. These cells are called **secondary cells** and the process is called **charging** or *recharging* the cell (battery). The cells and batteries used in mobile phones and cars are obvious examples of secondary cells. When energy is supplied to a current in a circuit, we say that the cell is **discharging**.
- Figure 5.31 shows a car battery with six secondary cells which are automatically recharged while the car engine is operational.

Figure 5.31

Expert tip

A great deal of on-going scientific research is taking place into the design of secondary cells. The aims are to (i) improve the *energy density*, increasing the energy that can be stored in each cm³, so that the size of the cells can be reduced; (ii) reduce the time needed for recharging; (iii) increase the number of times a cell can be recharged.

Key concept

The current needed to recharge a secondary cell must flow in the opposite direction to the current flow in normal use.

■ Identifying the direction of current flow required to recharge a cell

- Figure 5.32 shows how a recharging voltage needs to be connected. The magnitude of the recharging voltage is the same as the voltage provided by the cell in normal use.

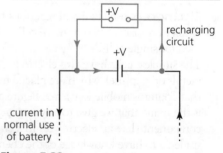

Figure 5.32

Describing the discharge characteristic of a simple cell (variation of terminal potential difference with time)

- When we use a battery we want the voltage that is supplies (its *terminal potential difference*) to stay constant for as long as possible. If, during use, the voltage decreased slowly, it would probably mean that battery became of no use while there was still a lot of energy stored inside it.
- Figure 5.33 shows (in blue) a typical **discharge characteristic** of a simple cell supplying a constant current. The red line shows the behaviour of an 'ideal' cell, the voltage of which only decreases when *all* of its stored energy had been transferred.

Key concept

The discharge characteristic of a simple cell shows that its terminal p.d. remains almost constant until most of its energy has been transferred.

Expert tip

The area under a discharge characteristic, for any chosen time *t* equals *Vt*, and since the total energy transferred in time *t* equals *VIt*, we can use the area to determine the energy transferred by a constant current.

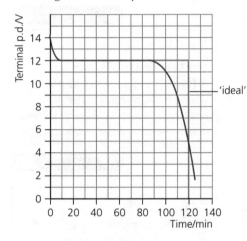

Figure 5.33

Investigating practical electric cells (both primary and secondary)

- The discharge characteristic of a store-bought cell is an experiment which may take many hours and is well suited to the collection of measurements using data-loggers and computer analysis. Similarly, the charging of a secondary cell can be investigated.
- There are a great number of variables that can be altered in laboratory constructed primary cells, so there are many possible investigations.

QUESTIONS TO CHECK UNDERSTANDING

48 A mobile phone battery was completely discharged when it was connected to a 4V 'charger'. When completely charged it stores 50kJ of chemical energy. Sketch a graph to show how the charging current might vary with time. Include estimates of numerical values.

49 The battery for a mobile phone is rated at 3.8V, 2800mAh.

   a Is it a primary or secondary cell?

   b How much energy (J) is stored in a charged battery?

   c If the chemicals inside the cell have a volume of 6.5cm³, what is their energy density when fully charged?

50 Estimate the energy stored in the battery whose discharge characteristic is shown in Figure 5.33. Assume that the discharge current was a constant 2.0A.

51 Outline an experiment to investigate the discharge characteristic of a store-bought 1.5V battery.

Internal resistance

- Figure 5.34 shows a common way of representing a cell with its internal resistance: as if they were separate components. But the point between the cell and the internal resistance is imaginary, it cannot be accessed.

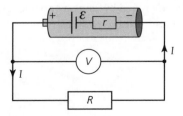

Figure 5.34

Key concept

Batteries and other sources of electrical power are not perfect conductors of electricity. They all have resistance, called their **internal resistance**, *r*.

- It may be thought desirable that the internal resistance of a power source is much less than the resistance of the rest of the circuit, it may then be described as *negligible* and therefore be ignored in any calculations. In examination questions it is common to assume that internal resistance is negligible *unless* it is mentioned (because that question is about internal resistance specifically).
- However, in some circuits, a value for the internal resistance of the source must be included in a calculation of the total resistance of the circuit (= $R + r$).
- In this course we will assume that the internal resistance of a source is constant.

Terminal potential difference

- Terminal potential difference can be measured directly and easily with a voltmeter connected in parallel with the terminals. It may not be constant. It is important to distinguish between terminal potential difference and *electromotive force (emf)*.

> **Key concept**
>
> The **terminal potential difference** of a battery or power source is the voltage across its terminals, which is also the voltage across the rest of the circuit.

Electromotive force (emf)

- The emf of a battery, for example, is the total energy per coulomb that can be transferred from chemical energy within it.
- The term *electromotive force* may be considered to be misleading because it is a potential difference *not* a force. Simply using the initials *emf* is recommended.
- We may refer to the emfs produced by any device which transfers another form of energy to electrical energy: cells, batteries, dynamos, generators, photovoltaic (solar) cells and microphones.
- When a current flows through an energy source which has internal resistance, some energy will be dissipated as internal energy *within the source*. This means that the terminal potential difference will be less than the emf. The difference is often called **lost volts**.
- The emf of a source cannot be measured directly with a voltmeter *unless* the internal resistance is negligible and/or a very high resistance voltmeter is used.
- The high-resistance voltmeter in the circuit shown in Figure 5.35 is measuring the emf of the cell when the switch is open. But when the switch is closed and a current flows, the voltmeter reading (which shows the terminal p.d.) will decrease because of the voltage 'lost' across the internal resistance (Ir).

> **Key concept**
>
> The **electromotive force (emf)**, ε, of a cell or any other source of electrical power, is defined as the *total* energy that can be transferred in the source per coulomb passing through it.

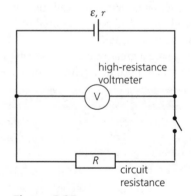

Figure 5.35

▇ Solving problems involving emf, internal resistance and other electrical quantities

- emf = terminal potential difference + lost volts
- Terminal potential difference $V = IR$; lost volts = Ir
- $\varepsilon = IR + Ir = I(R + r)$

▇ Determining internal resistance experimentally

- Internal resistance and emf of a cell can be determined experimentally by using a high-resistance voltmeter to measure the voltage across the cell when there is no current flowing (= ε), measuring the current, I, through a known resistor, R, and substituting into the equation above to determine the internal resistance.
- The accuracy of the experiment would be improved by using a variable resistor and collecting a range of different current and voltage readings.

> **Key concept**
>
> The size of the lost volts (Ir) increases with the size of the current in the circuit. That is, when a greater current is taken from a source which has significant internal resistance, the useful terminal p.d. falls; it is not constant.

QUESTIONS TO CHECK UNDERSTANDING

52 A battery of emf 1.5V and internal resistance 0.5Ω is connected to a 5.0Ω resistor.

   a What current flows in the circuit?

   b What reading will be seen on a voltmeter connected across the battery?

53 Explain why the headlights of a car may go slightly dimmer when the engine is started.

54 When a battery was connected to a resistor of 4.0Ω the current was 1.0A, but the current fell to 0.50A when the resistance was increased to 10Ω. Determine the emf and internal resistance of the battery.

55 Explain why a high-voltage supply for use in school demonstrations has a very large internal resistance included inside it.

56 Three cells, each of emf 1.6V and internal resistance 0.4Ω, are to be combined to make a battery. What are the four possible arrangements, and their overall emfs and internal resistances?

57 a Describe an experiment using a variable resistor to determine the internal resistance and emf of a shop-bought battery. Explain what readings you would take, how you would represent them graphically and how you would use the graph to get the results.

   b How would you expect the experimental value of the emf to compare to the value written on the side of the battery?

> **Expert tip**
>
> As well as the emf, the internal resistance of a battery is an important factor that has to be considered when a choice is made of which battery to use in a particular device. For example, the maximum power is transferred from a battery to a circuit when $r = R$.

NATURE OF SCIENCE

◼ Peer review

The growth in the use of secondary cells (and the devices they power, like smart phones and tablets) in recent years has been much greater than would have been generally predicted some years ago. At the same time there have been increasing environmental pollution issues concerning the disposal of electronic devices and their batteries. Society has to decide whether the risks of new technologies are worth accepting for the benefits gained. But that assumes, naively, that benefits and risks can be assessed in advance of new discoveries and inventions.

5.4 Magnetic effects of electric currents

Essential idea: The effect scientists call magnetism arises when one charge moves in the vicinity of another moving charge.

Revised ▢

Magnetic fields

Revised ▢

- We say that there is a **magnetic field** around a *bar magnet* because in that space another magnet or a magnetic material would experience a **magnetic force**. This is similar in principle to the gravitational fields around masses, and the electric fields around charges.

- However, we need to know more about the origins of magnetism before we can define magnetic field more generally. All magnetic fields are produced by *moving* charges. (There are *electric* fields around all charges, whether they are moving or not.) A flow of charge is a current, so *all* currents are surrounded by magnetic fields (and electric fields).

- A few materials have **permanent magnetic properties** because of the way in which charges (electrons) move inside atoms. The magnets we see in everyday life are made from such materials. A full explanation of permanent magnetism is a complex topic and it is *not* part of the IB Physics course.

> **Key concept**
>
> A magnetic field is the a region around moving electric charge (current) in which another moving charge (current) would experience a force.

Magnetic poles

Revised ▢

- Magnetic forces may be attractive or repulsive (like electric forces, but unlike gravitational forces). This means that we must identify two different 'types' of magnetism, although they always occur in pairs. We call them *north poles* and *south poles* (compare with *positive* and *negative* for the two types of charge).

- A simple bar magnet has a north pole at one end and a south pole at the other. They got these names because a freely supported bar magnet will twist until it is pointing N↔S. The end closer to the geographic north is called the north (seeking) pole, the other end is called the south (seeking) pole.
- This property has made freely supported bar magnets very useful for determining geographic directions. As such they are examples of **compasses**.
- The force between similar poles (N and N, or S and S) is repulsive. The force between opposite poles (N and S) is attractive. If two bar magnets are brought close to each other and at least one is free to rotate, the forces will result in the magnets aligning.
- The Earth has a magnetic field which behaves like an enormous bar magnet with a *magnetic south pole* near the *geographic north pole*. This is why the magnetic north pole of a compass needle always points to geographic north. See Figure 5.36.

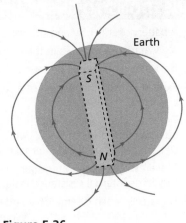

Figure 5.36

Permanent magnetic field patterns

Revised

- Magnetic fields are represented in drawings by field lines. Electric and gravitational fields are represented in similar ways.
- In any given field pattern, the field is strongest where the lines are closest. Field lines can never cross each other.

■ Sketching and interpreting magnetic field patterns

- The basic magnetic field pattern around a bar magnet is shown in Figure 5.37, which includes a small *plotting* compass indicating the direction of the field at one point.
- The field between bar magnets is shown in Figure 5.38.
- In each of these three diagrams it has been assumed that the magnets' fields are much stronger than the ever-present magnetic field of the Earth (which has been ignored).

> **Key concept**
> A magnetic field line shows the direction of force that would be exerted on a (hypothetical) single north pole placed at that point (from magnetic north to south, the same as a compass points).

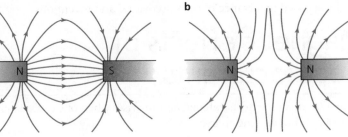

Figure 5.37   **Figure 5.38**

Magnetic field patterns around currents

- The magnetic fields produced by currents have a very wide range of applications, including electromagnets, generators, motors, and transformers.
- The basic field (shown in the diagram on the left-hand side of Figure 5.39) is produced by a direct current flowing in a long straight wire.
- The field is circular and the separation of the lines increases with distance from the wire because the field is getting weaker.
- Increasing the current increases the strength of the magnetic field. Reversing the direction of the current reverses the direction of the field. We say that the **polarity** has been reversed.

■ Determining the direction of the magnetic field based on current direction

- Although the field is circular, it still has direction. The direction can be determined using the '**right-hand grip rule**' as shown in the right-hand side of Figure 5.39: if the thumb points in the direction of the current, the fingers indicate the direction of the field.

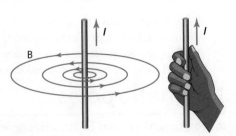

Figure 5.39

◼ Magnetic fields around currents in coils and solenoids

- ◼ The easiest way of producing a stronger field is by wrapping the wire round and round to make a coil. Each turn adds approximately equally to the strength of the field.

- ◼ A coil which is made by wrapping turns of insulated wire closely together, but without overlapping them, is called a **solenoid**. Solenoids are very useful because of the *strong uniform* magnetic fields that can be produced inside them.

- ◼ Figure 5.40 represents the magnetic field in and around the current in a solenoid (there are usually many more turns than shown in this figure). The polarity can be determined by careful use of the 'right-hand rule'. Reversing the current reverses the poles. The strength of the field depends on the number of turns per cm and the current.

- ◼ **Soft iron** has the very important advantage of gaining and losing its magnetism very quickly. This makes it ideal for electromagnetic devices (e.g. electromagnets).

- ◼ A steel core also has good magnetic properties but it tends to retain its magnetism after the current has been switched off. This means that the magnetism is not easily controlled.

- ◼ The **permeability of free space** is an important fundamental constant, which is given the symbol μ_0. Its value is $4\pi \times 10^{-7}\,\text{T m A}^{-1}$. All other materials have magnetic permeabilities with higher values.

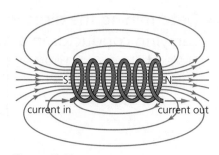

Figure 5.40

> ### Key concepts
>
> Strong magnetic fields are produced inside coils and solenoids carrying direct currents. Placing a core made of soft (pure) iron inside will make the field much stronger.
>
> **Permeability** represents the magnetic properties of a medium.

QUESTIONS TO CHECK UNDERSTANDING

58 a Sketch the magnetic field pattern in a plane perpendicular to a wire carrying a current perpendicularly into the paper you are drawing on.

   b Sketch the magnetic field pattern in the room where you are reading this. Include a small circle with an arrow inside to represent the direction in which a compass would point.

59 a Make a copy of Figure 5.41, which shows how a strong uniform magnetic field can be produced between the poles of a U-shaped soft iron core.

   a Label the north and south poles.

   b Draw field lines to represent the uniform field produced.

   c Give two reasons why soft iron is used for the core of this electromagnet.

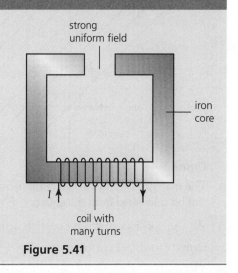

Figure 5.41

Magnetic force

Revised ▢

- ◼ When a current flows across a magnetic field, the field due to the current and the external field combine to produce a force on the current. The force is perpendicular to both the plane that contains the field and the direction of the current. See Figure 5.42. This is often called the **motor effect**.

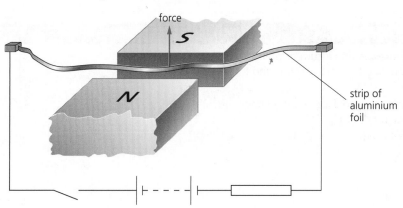

Figure 5.42

Determining the direction of force on a current-carrying conductor in a magnetic field

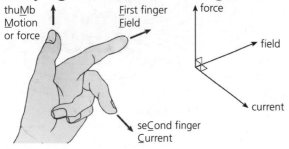

Figure 5.43

> **Key concepts**
>
> The '**left-hand rule**' can be used to predict the relative directions of magnetic field, current and force. See Figure 5.43.
>
> The same rule can be used to determine the direction of forces on individual charges.
>
> The magnitude of the force can be determined from the equation $F = BIL \sin \theta$.

- The size of the magnetic force, F, depends on the strength of the magnetic field, B, the magnitude of the current, I, the length of the conductor in the field, L and the angle between the current and the field, θ. See Figure 5.44, in which the force produced would be perpendicularly out of the page.

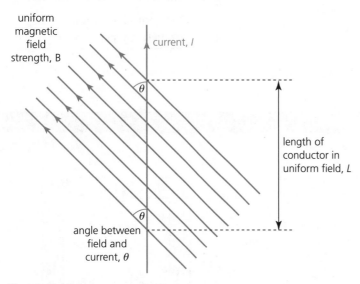

Figure 5.44

- The magnitude of force on a current-carrying conductor in a magnetic field can be calculated from the equation $F = BIL \sin \theta$.

- We can consider that this equation defines **magnetic field strength**, B, in terms of force per current metre: $B = \dfrac{F}{IL \sin \theta}$

- The units of *magnetic field strength*, as can be seen from the equation, are $N\,A^{-1}\,m^{-1}$, more commonly called **teslas**, T.

- The effect shown in Figure 5.42 is commonly known at the *motor effect*, because magnetic forces can be used to produce continuous rotation of a current-carrying coil in a magnetic field (details are not required for this course).

Solving problems involving magnetic forces, fields and current

QUESTIONS TO CHECK UNDERSTANDING

60 A force of 4.3×10^{-4} N was exerted on a wire carrying a current of 3.2 A across a magnetic field. The angle between the wire and the field was 60° and the length of the wire in the field was 4.8 cm.

   a Sketch the arrangement of the wire and the field, labelling the known quantities and the direction of the magnetic force produced.

   b Determine the strength of the magnetic field.

61 Figure 5.45 represents currents flowing in opposite directions in two parallel wires.

a Copy the diagram and sketch the field lines around wire Y.

b Mark on wire X the direction of the force exerted on the current due to the current being in the field of Y.

c Deduce the direction of the force on wire Y.

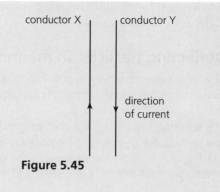

Figure 5.45

62 A square coil of wire carrying a current of 0.43 A rotates in a uniform magnetic field of strength 0.67 T. See Figure 5.46. If the coil has 500 turns (the figure only shows one turn), and the coil has sides of length 2.4 cm, calculate the maximum force exerted on one side of the coil.

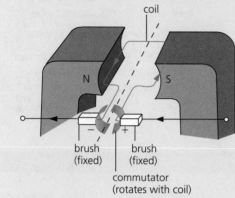

Figure 5.46

Magnetic forces on individual moving charges

- The equation $F = BIL \sin \theta$ involves macroscopic measurements concerning the force on a large number of moving charges in a current, but it can also be applied to individual charges. See Figure 5.47, which shows a single charge, q, moving with a velocity v across a magnetic field.

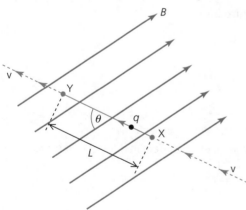

Figure 5.47

- If the charge moves a distance L in time t, velocity, $v = L/t$. We also know that $I = q/t$. These equations can be combined with $F = BIL \sin \theta$ to give an equation for the force on a single charge moving across a magnetic field:
$F = qvB \sin \theta$

Determining the direction of force on a charge moving in a magnetic field

- The magnetic force is perpendicular to the field and the velocity of the particle.
- We can use the *left-hand rule* to predict the direction of the magnetic force on moving charges, remembering that the direction of current flow is always that

> **Key concept**
>
> If beams of charged particles are directed perpendiculary across magnetic fields, they will move along the arcs of circles. The radii of their paths can be used to determine properties of the particles.

of moving *positive* charge. This will be in the opposite direction to the motion of electrons, or negative ions.

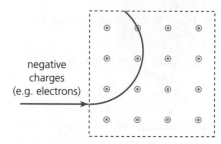

negative charges (e.g. electrons)

Figure 5.48

■ Deflecting particles in magnetic fields

- The most common situation is one in which charged particles (in a *particle beam*) are directed to move perpendicular to a magnetic field (so that sin θ = 1).
- The resulting force which is always perpendicular to motion will provide the centripetal force to make the particles move in the arc of a circle (see Section 6.1). Figure 5.48 shows the example of the deflection of negative charges moving perpendicular to a magnetic field directed out of the paper.
- If θ is between 0° and 90° the particles will move in a helical path.
- The magnetic deflection of charged particle beams has some important applications, including high-energy nuclear physics accelerators and mass spectrometers for identifying isotopes (no details required). Detecting the motion of charged particle beams in magnetic and electric fields is often the best way of determining the properties of the particles.

> **Expert tip**
>
> In Section 6.1 the equation for any centripetal force is shown to be
> $F = \dfrac{mv^2}{r}$. When compared to
> $F = qvB$, (sin θ = 1), it can be shown that the radius of the path of a charge moving perpendicularly across a magnetic field is given by $r = \dfrac{mv}{qB}$.

■ Solving problems involving magnetic forces, fields and charges

QUESTIONS TO CHECK UNDERSTANDING

63 Figure 5.49 shows the deflection of charged particles (ions) passing perpendicularly across a magnetic field of strength 0.12 T.

  a   Are the particles positively or negatively charged?

  b   Explain why the paths are arcs of circles.

  c   Determine the force on the charges if the ions were all singly charged and all travelled at a speed of $4.4 \times 10^6\,\mathrm{m\,s^{-1}}$.

  d   Suggest a possible reason why three different paths are seen.

64 Using knowledge from Chapter 7 on nuclear radiations, qualitatively compare the deflections of alpha particles, negative beta particles and gamma rays as they pass through the same magnetic field.

65 Explain how it would be possible for an electron beam to pass through a magnetic field without having its velocity changed, but will always be affected by an electric field.

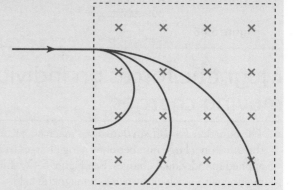

Figure 5.49

66 A beam of electrons is directed at an angle of 45° across a magnetic field of strength 38 mT.

  a   Sketch this arrangement, including arrows to represent directions.

  b   In which direction will the beam be deflected?

  c   If the force on the individual electrons was $7.8 \times 10^{-14}\,\mathrm{N}$, what was the speed of the electrons?

67 Use the equation $r = mv/qB$ to determine the strength of a perpendicular magnetic field required to make a hydrogen ion (proton) travelling at 10% of the speed of light move in a circle of radius 1.0 m.

NATURE OF SCIENCE

■ Models and visualization

One of the reasons that the study of fields is a difficult concept in physics is that they cannot be seen, although any empty space with a field is fundamentally different from a space without it. 'Lines of force' were first used in the early nineteenth century and the 'simple' representation of fields with lines is a surprisingly powerful visualization tool, providing a mental picture (model) that is useful for both students and experienced scientists.

6 Circular motion and gravitation

6.1 Circular motion

Essential idea: A force applied perpendicular to its displacement can result in circular motion.

- Many objects move in circles, or paths which follow the arcs of circles. In particular, Section 6.2 concentrates on the motion of planets and satellites.
- If an object spins on its own axis we describe it as a *rotation*. If an object moves around a separate centre we describe it as a *revolution*. The Earth revolves around the Sun and at the same time rotates on its own axis.
- Before we can develop an understanding of circular motion, we need to be clear about the measurement of angles.

Degrees and radians

- The use of *degrees* to measure angles is familiar to us in everyday life and in our early studies in mathematics and science, but it is not convenient mathematically because it is based on an arbitrary and historical choice of 360° for a complete revolution. The ratio of the circumference of a circle, $2\pi r$, to its radius, r, is a more logical basis for the measurement of angles. In this system $\frac{2\pi r}{r}$ ($= 2\pi$) becomes the measurement of one revolution, rather than 360°.
- Although an angle like 2π is just a ratio, it is called 2π **radians**, commonly reduced to rad.
- $180° = \pi$ rad. 1 rad $= \frac{180°}{\pi} = 57.3°$.
- In general, an angle in radians is equal to the distance along the arc of a circle divided by the radius, θ (rad) $= \frac{s}{r}$ in Figure 6.1a.

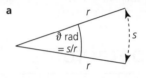

a

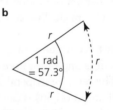

b

Figure 6.1

Period, frequency, angular displacement and angular velocity

- For the common situation of an object moving continuously and uniformly in circles, the angular velocity and the linear speed of the object around the circumference will both be constant, and they are closely related.
- For continuous uniform circular motion, $\omega = \frac{2\pi}{T} = 2\pi f$.
- The linear speed of an object moving uniformly in a circle, $v = \frac{2\pi r}{T}$.
- Therefore, angular velocity, ω, and linear speed, v, are related by the simple equation $v = \omega r$. See Figure 6.2.

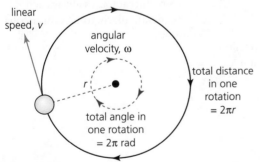

Figure 6.2

> ### Key concepts
>
> The **period**, *T*, of a circular motion is defined as the time taken for the object to complete one rotation (2π radians or 360°).
>
> The **frequency**, *f*, of circular motion is defined as the number of rotations in unit time. $f = \frac{1}{T}$ (unit: Hz).
>
> **Angular displacement**, *θ*, is defined as the angle through which a rigid object has rotated from a fixed reference position.
>
> **Angular velocity,**
> $\omega = \frac{\text{angular displacement}}{\text{time}}$; $\omega = \frac{\Delta\theta}{\Delta t}$. It has the unit rad s^{-1}.

■ Solving problems involving period, frequency, angular displacement, linear speed and angular velocity

Centripetal force

- From Newton's first law of motion (Section 2.2), we know that any object that is not moving in a straight line at a constant speed must have a resultant force acting on it.
- We say that any force which is producing circular motion is acting as a *centripetal force*. It is important to realize that centripetal force is not another type of force. The term 'centripetal' simply means that the force being discussed (of whatever origin) is causing circular motion.
- Any centripetal force must continuously change direction so that it is always directed towards the centre of the circle. See Figure 6.3.

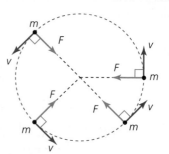

Figure 6.3

- Centripetal forces can have a variety of different origins, for example, tension, friction, gravitational, electrical, magnetic or reaction forces. Some examples are provided below.

■ Identifying the forces providing the centripetal forces such as tension, friction, gravitational, electrical, or magnetic

- *Tension* provides the centripetal force on an object which is spun around (almost) horizontally on the end of a length of string.
- *Friction* provides the centripetal force for people or cars moving around corners.

- *Gravitational forces* provide the centripetal forces for satellites moving around the Earth, or planets moving around the Sun.
- *Electrical forces* provide the centripetal forces that keep electrons moving around nuclei in atoms.
- *Magnetic forces* provide the centripetal forces which maintain particle beams in circular paths in nuclear accelerators (see Section 5.4).
- *Reaction forces* are also commonly involved with circular motion. As examples consider (i) water in a bucket being whirled in a vertical circle (consider the water, not the bucket); (ii) a cyclist moving at speed around a *banked* track. Friction between the track and the tyres is not enough to provide sufficient centripetal force at the high speed. See Figure 6.4, in which the normal reaction force on the tyre has a horizontal component acting towards the centre of the circle.

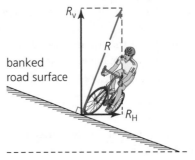

Figure 6.4

QUESTIONS TO CHECK UNDERSTANDING

5   Look at Figure 6.5. The girl on the swing is moving along the arc of a circle.

  a   What is providing the centripetal force on: **i** the swing seat, **ii** the girl?

  b   Draw a free-body diagram to show the forces acting on the girl (represented as a point mass).

Figure 6.5

6   What provides the centripetal force when a plane changes direction (at a constant altitude)?

7   See Figure 6.6.

  a   What provides the centripetal force for these athletes running on a curved track?

  b   What design features make sure that this force can be large enough for high speeds?

Figure 6.6

Centripetal acceleration

- Any object which is moving in a circular path has a continuously changing velocity, even if it has a constant speed (because velocity is a vector quantity). This means that there must be an acceleration which is directed in the same direction as the force, towards the centre of the circle.
- Centripetal forces produce *centripetal accelerations*. See Figure 6.7.

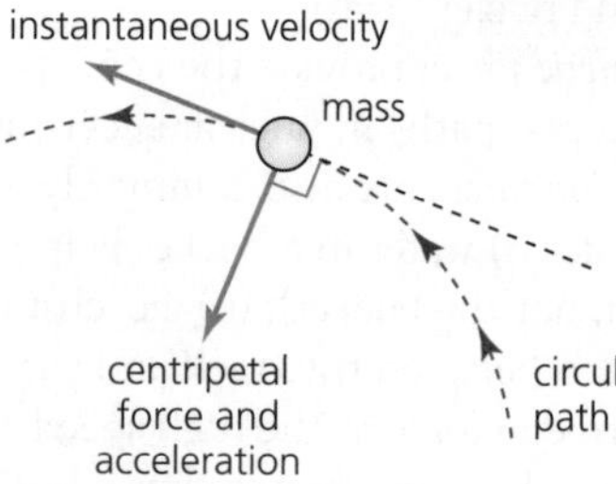

> **Key concept**
>
> Any object moving along a circular path has a **centripetal acceleration**, a, towards the centre of the circle.
>
> $a = \frac{v^2}{r}$

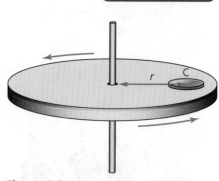

Figure 6.7

- Since $v = \omega r$, the equation for centripetal acceleration can also be written as $a = \omega^2 r$.
- Since $v = \frac{2\pi r}{T}$ the equation can also be written as: $a = \frac{4\pi^2 r}{T^2}$.

Calculating centripetal force

- Newtons second law (Section 2.2), $F = ma$, can be used to obtain an equation for centripetal force.
- This equation enables us to calculate the theoretical force needed for a particular circular motion. The actual force acting may, or may not, be able to provide a force of the required magnitude. For example, we could use the equation above to calculate the centripetal force needed for a car to go around a corner at a particular speed, but in practice the actual friction between the road and the tyres might not be enough.

> **Key concept**
>
> The centriptal force needed to keep a mass m moving with a speed v in a circle of radius r can be calculated from
>
> $F (= ma) = \frac{mv^2}{r} = m\omega^2 r$.

Qualitatively and quantitatively describing examples of circular motion including cases of vertical and horizontal circular motion

- Consider Figure 6.8, which shows a coin, C, of mass 24 g, on a rotating turntable. The force of friction provides the necessary centripetal force and the coin stays in the same place on the turntable. If it completes 5 rotations in 9.8 s the angular velocity, ω, is $\frac{10\pi}{9.8} = 3.2\,\text{rad}\,\text{s}^{-1}$. If the radius is 8.3 cm, the centripetal acceleration is $\omega^2 r = 0.85\,\text{m}\,\text{s}^{-2}$ and the force of friction providing the centripetal force is $m\omega^2 r = 0.020\,\text{N}$. If the speed of rotation is increased, the centripetal force needed to keep the coin in place will also increase. If friction cannot provide this force, the coin will slip off the turntable along a tangential path.
- When water in a bucket is whirled in a vertical circle, the reaction force from the base of the bucket helps to provide the necessary centripetal force on the water. See Figure 6.9. If the angular velocity is great enough, the water will remain in the bucket, but the water will lose contact with the base of the bucket (at its highest point) if its weight is greater than the centripetal force needed at that velocity. That is, if $mg > m\omega^2 r$. For example, if the radius of the circle is 1 m and one revolution is completed every 2 s, the water will just stay in the bucket.

Figure 6.8

bucket of water

■ Solving problems involving centripetal force, centripetal acceleration, period, linear speed and angular velocity

Figure 6.9

QUESTIONS TO CHECK UNDERSTANDING

8   A small rubber ball on the end of a thin string was spun in vertical circles.

   a   What provided the centripetal force on the ball?

   b   Draw a free-body diagram to show the forces acting on the ball when the string was horizontal.

   c   The speed of rotation was increased until the string broke. Explain why the string was most likely to break when the ball was at the lowest point.

   d   In which direction will the ball then move?

9   A car of mass 1400 kg moves with a constant linear speed of 12 m s^{-1} around a bend in a horizontal road which is the arc of a circle of radius 200 m.

   a   Calculate the centripetal acceleration of the car.

   b   What is the magnitude of the centripetal force on the car?

   c   Determine a value for the coefficient of dynamic friction between the tyres and the road.

10   The Earth moves around the Sun in an orbit with an average radius of 150 million kilometres.

   a   What provides the centripetal force on the Earth?

   b   The same magnitude force acts on the Sun. Suggest its effect.

   c   What is the angular velocity of the Earth around the Sun?

   d   Calculate the magnitude of the centripetal force acting on the Earth. (Mass of the Sun = 2.0 × 10^{30} kg; mass of the Earth = 6.0 × 10^{24} kg.)

11   An alpha particle beam has particles with masses of 6.6 ×10^{-27} kg and speeds of 1.5 × 10^{7} m s^{-1}.

   a   How can the beam be made to move in a circular path?

   b   What magnitude centripetal force on each particle will make the beam move in a circular path of radius 200 cm?

   c   Use the equation $F = qvB \sin \theta$ (Section 5.4) to determine the strength of the magnetic field that produces this path.

12   A satellite orbits at a distance of 12 750 km from the Earth's centre. At this height the acceleration due to gravity, g, is 2.45 m s^{-2}. Calculate the period of the satellite and how many times it orbits the Earth daily.

NATURE OF SCIENCE

■ Observable universe

Section 6.1 is another example which shows how a vast range of different common observations can, given sufficient time and insight, be reduced to having a common explanation, in this case they are all based on the concept of centripetal force.

It is a basic assumption of science that all our observations can, in principle, be explained. Scientists generally do not believe that there are things beyond reason, even if there are always some phenomena which are currently unexplained.

6.2 Newton's law of gravitation

Revised ☐

Essential idea: The Newtonian idea of gravitational force acting between two spherical bodies and the laws of mechanics create a model that can be used to calculate the motion of planets.

■   Attractive **gravitational forces** exist between all masses. The same force acts on both masses (even if their masses are different). See Figure 6.10, which shows the forces on two masses, M and m, separated by a distance r. (Sometimes m_1 and m_2 are used instead.)

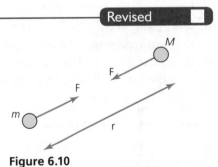

Figure 6.10

Newton's law of gravitation

Revised ☐

■   The magnitude of the gravitational force between two isolated *point masses* is proportional to the product of the masses and inversely proportional to the square of the distance between them. This is similar in principle and mathematical form to Coulomb's law (Section 5.1). A **point mass** is a theoretical concept used to simplify the discussion of forces acting on objects.

> **Key concept**
>
> Newton's universal law of gravitation: $F = \frac{GMm}{r^2}$, where
>
> G is the **universal gravitation constant**, which has a value of $6.67 \times 10^{-11}\ \text{N m}^2\,\text{kg}^{-2}$.

- This law applies to *point masses*, but it can also be used *outside* of large *spherical masses*, like planets (of evenly distributed density), which have gravitational fields that behave as if all their masses were concentrated at their centres. r is then the separation of the *centres* of the two masses concerned.

Expert tip

There is a very strong *analogy* between Coulomb's law and Newton's law. The similar mathematical equations both represent fundamental forces which follow an inverse square law relationship with distance. Understanding one of these laws greatly helps you to understand the other. The magnitudes of the two constants involved (k and G) are very different and this reflects the fact that electrical forces are much, much greater than gravitational forces (under similar circumstances).

Gravitational fields

Revised ☐

- All masses create gravitational fields around themselves, but a gravitational field is negligible unless the mass is very large (like the Earth).

■ Gravitational field strength

- Gravitational field strength can be calculated from $g = \frac{F}{m}$ (unit: N kg^{-1}, but m s^{-2} is equivalent).
- The "test mass" involved in the definition of gravitational field strength must be so small that it does not affect the field.
- The gravitational field strength on the surface of the Earth is easily determined by weighing a known (test) mass.

> **Key concepts**
>
> A **gravitational field** exists anywhere where a mass experiences a gravitational force.
>
> **Gravitational field strength**, g, at a point is defined as the force per unit mass experienced by a small test mass at that point.

Common mistake

It is easy to confuse G and g! G is one of the few universal constants in science and, as far as we know, it has exactly the same value everywhere and for all time. However, g, the strength of a gravitational field, is *not* a constant. It varies with location, although we may loosely describe its value at all points on the Earth's surface as constant.

- The definition of gravitational field can be compared to similar ones for electric and magnetic fields.
- Gravitational field strength is a vector quantity, pointing in the direction of the force.
- The gravitational force on a mass near to a planet is called its *weight*. The same force acts in the opposite direction on the planet.
- Since $g = \frac{F}{m}$, and $F = \frac{GMm}{r^2}$, the gravitational field strength at a distance r from the centre of a planet of mass M is given by $g = F = \frac{GM}{r^2}$.
- Figure 6.11 illustrates how the gravitational field strength varies with distance from the Earth, following an *inverse square law*.

Expert tip

Newton calculated the centripetal acceleration of the Moon (from $a = \frac{v^2}{r}$) to show that it gave the same value as $\frac{g}{60^2}$, as shown in Figure 6.11. This was conclusive evidence that the inverse square relationship was correct.

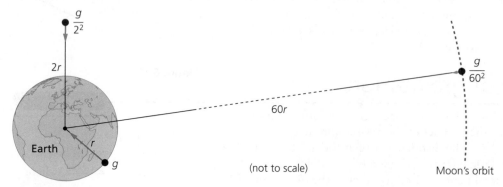

Figure 6.11

- Figure 6.12 shows the inverse square variation of gravitational field strength, *g*, with distance, *r*, from the centre of a planet. r_p represents the radius of the planet. The field strength is greatest on the surface and reduces towards the centre because of the way in which the mass of the planet is distributed around the point in question.

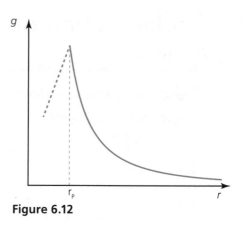

Figure 6.12

■ Determining the resultant gravitational field strength due to two bodies

- Sometimes an object may be in two or more significant gravitational fields, for example between a planet and a moon. The combined field, in magnitude and direction, can be calculated using *vector addition* of the individual fields.
- Figure 6.13 illustrates how the resultant gravitational field varies between the Earth and the Moon.

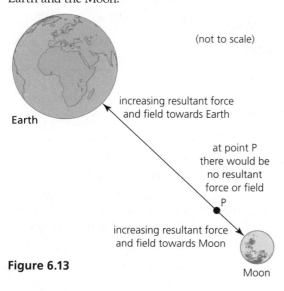

(not to scale)

Earth

increasing resultant force and field towards Earth

at point P there would be no resultant force or field

P

increasing resultant force and field towards Moon

Moon

Figure 6.13

> **Key concept**
> Vector addition (Section 1.3) is needed to determine the resultant of two gravitational field strengths.

QUESTIONS TO CHECK UNDERSTANDING

13 a Calculate the gravitational forces between the Earth and the Moon. (Mass of Earth is 5.97×10^{24} kg and the mass of Moon is 7.35×10^{22} kg. Their centres are an average 3.84×10^5 km apart.)

b This force keeps the Moon in orbit around the Earth, but does it have any noticeable effect on the Earth?

14 Calculate an order of magnitude for the gravitational forces acting between two cars parked 2 m apart.

15 Explain why it is very difficult to measure gravitational forces (except for weights).

16 The planet Venus has a mass of 4.9×10^{24} kg and a radius of 6052 km. Determine the gravitational field strength:

a on its surface

b 2000 km above its surface.

17 a At what distance from the Earth's centre is the gravitational field strength 1.0 N kg^{-1}? (Radius of Earth is 6.4×10^6 m.)

b What is the field strength at the altitude of a satellite which is 300 km above the Earth's surface?

18 Figure 6.14 shows a planet of mass 8.3×10^{24} kg and its moon (not to scale). At the point P the combined gravitational field strength is zero.

a What is the gravitational field strength at P due to only the planet?

b Determine the mass of the moon.

c What is the combined field at point Q, which is the same distance from the Moon as point P?

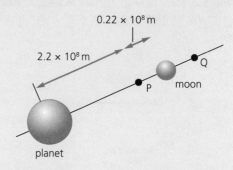

0.22×10^8 m

2.2×10^8 m

P

Q

moon

planet

Figure 6.14

Circular motion of planets, moons and satellites

■ Describing the relationship between gravitational force and centripetal force

■ The **orbits** of the planets around the Sun, and moons and satellites around planets, may be considered to be circular (although this is not necessarily true, many are elliptical).

■ A **satellite** is an object which moves in a curved path (orbits) around a much larger mass. Satellites may be natural (like the Earth or the Moon), or artificial (as used for communication, for example).

■ Applying Newton's law of gravitation to the motion of an object in circular orbit around a point mass

■ The equation in the 'Key concept' box reduces to $v^2 = \dfrac{GM}{r}$.

■ Since we have seen that field strength $g = \dfrac{GM}{r^2}$, we can re-write the previous equation as $v^2 = gr$, showing directly how the linear speed depends on the field strength.

■ Clearly, the speed and period of an orbiting object do *not* depend on the mass of the object (for the same reason that projectiles of different masses follow the same trajectories: see Section 2.1). For example, all satellites in circular orbits at the same height will take the same time to orbit the Earth.

■ The period of any orbit, T, can be determined by substituting $v = \dfrac{2\pi r}{T}$ into the previous equations.

■ The equations above show us that any artificial satellite in a circular orbit must be given the right orbital speed, v, for its required distance from the centre of the Earth.

> **Key concept**
>
> Gravity provides the centripetal force for the orbits of moons, satellites and planets. So that,
> $F = \dfrac{mv^2}{r} = \dfrac{GMm}{r^2}$ (assuming that the object is acting like a point mass and the orbit is circular).

> **Key concept**
>
> The linear speed, v, of any mass in a circular orbit depends only on the mass of the central object, M, and the separation of their centres, r.
> $\left(v^2 = \dfrac{GM}{r}\right)$

Expert tips

The relationship between period and radius of planets orbiting the Sun was first identified by Johannes Kepler. Kepler's third law: $\dfrac{r^3}{T^2} =$ constant (shown by Newton to be equal to $\dfrac{GM}{4\pi^2}$ by combining $v^2 = \dfrac{GM}{r}$ with $v = \dfrac{2\pi r}{T}$).

We are aware of our own weight because of the reaction force from the ground pushing up on us. If the ground suddenly disappeared we would accelerate downwards and feel 'weightless', even though our weight would not have changed. Similar sensations can be felt on some fairground rides. For the same reasons, any person in or on a container/vehicle which is in 'free-fall' (gravity is the only force acting), such as in a satellite orbiting the Earth, can lose contact with the surrounding surfaces and *appear* weightless.

■ Solving problems involving gravitational force, gravitational field strength, orbital speed and orbital period

QUESTIONS TO CHECK UNDERSTANDING

19 The mass of the Sun is 2.0×10^{30} kg.

  a Determine the gravitational field strength of the Sun at a distance of 150 million kilometres from its centre.

  b This distance is the average radius of the Earth's orbit. Use $v^2 = \frac{g}{r}$ to calculate the linear speed of the Earth around the Sun.

  c Show that the period of its orbit is one year.

20 Two planets orbit the same star at distances of 4.8×10^{10} m and 7.9×10^{11} m. If the first planet has a period of 200 Earth days, what is the period of the second?

21 What is the required orbital speed for a satellite designed to circle the Earth at a height of 1000 km?

22 Ganymede, with a mass of 1.48×10^{23} kg and a radius 2630 km is the largest moon in the solar system. It orbits Jupiter and the average separation of their centres is 1.07×10^6 km.

  a If Ganymede has an orbit speed of $10.9 \,\mathrm{km\,s^{-1}}$, what is its orbital period?

  b What is the gravitational field strength of Jupiter at the height of Ganymede's orbit?

  c What is the gravitational field strength of Ganymede on its surface?

23 Calculate the distance from the centre of the Earth to a satellite which has a period of 24 h.

NATURE OF SCIENCE

■ Laws

Newton's law of gravitation and the laws of mechanics are the foundation for deterministic classical physics. These can be used to make predictions but do not explain why the observed phenomena exist.

7 Atomic, nuclear and particle physics

7.1 Discrete energy and radioactivity

Essential idea: In the microscopic world energy is discrete.

Spectra

- **Spectra** are caused by energy changes inside atoms, so the production and analysis of spectra (**spectroscopy**) plays an important part in atomic physics.
- In Section 4.4 we saw that a beam of electromagnetic waves (light), containing different wavelengths can be dispersed into a spectrum by using a *prism*. (In Chapter 9, for HL students, we see that *diffraction gratings* can also be used to produce spectra.) The visible spectrum of white light is shown in Figure 7.1.

- Figure 7.1 is an example of a **continuous spectrum** in which the colours merge into each other and there are no gaps (or lines). Continuous spectra are typically produced by very hot objects, for example, the light *emitted* from a filament bulb or from the Sun forms a continuous spectrum. Analysis of a simple continuous spectrum does *not* provide any information about the nature of the source of the radiation.
- However, analysis of **line spectra** leads to a great deal of very useful information about energy levels within atoms and molecules. There are two kinds of line spectra: *emission spectra* and *absorption spectra*.

Figure 7.1

> ### Key concepts
> Line spectra are evidence of discrete energy levels within atoms.
>
> An **emission spectrum** is produced when *excited* atoms (or ions or molecules), usually in a gas at low pressure, emit electromagnetic radiation of precise wavelengths as excited electrons return to lower energy levels.

■ Describing the emission and absorption spectra of common gases

- They are called *line spectra* because, after the radiation has been made to pass through one or more narrow slits placed close to the source and then through a prism or grating, the observer sees a series of short vertical lines. See Figure 7.2, which shows prisms being used to produce an emission spectrum and an absorption spectrum of the same gas. Atoms and molecules in gases can be 'excited' by using high voltages (or high temperatures).

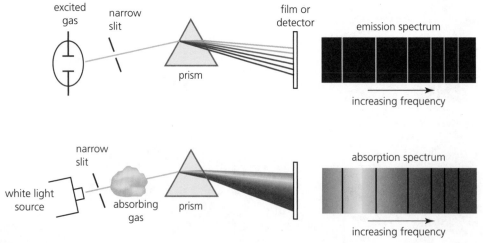

Figure 7.2

- An *emission spectrum* consists of a series of coloured lines, on a black background. Each spectrum is characteristic and unique to the element or compound that produced it. In this way, a line spectrum is a useful way of identifying unknown substances.

- An *absorption spectrum* consists of a series of black lines on a background of an otherwise continuous spectrum. The pattern of lines is the same as from the emission spectrum of the same element or compound (because the energy transitions are the same).
- After energy from a beam of radiation has been absorbed by the atoms, it is quickly re-emitted, but randomly in different directions. This results in a significant drop in intensity for radiation travelling in the original direction.
- Accurate measurements taken from experiments such as those shown in Figure 7.2 can be used to calculate the wavelengths and/or frequencies of the radiation that resulted in the lines. This is covered in Section 9.3 (HL).
- The fact that radiation energy can only be emitted or absorbed from atoms at certain frequencies is very important because it suggests that these processes involve movement between definite **atomic energy levels**. (If all the frequencies of a continuous spectrum were observed, it would suggest that atoms could move from *any* energy to any other energy.)

> **Key concept**
>
> An **absorption spectrum** is produced when some wavelengths of broad-band ('white') electromagnetic radiation are *absorbed* by gaseous atoms (or ions or molecules) at low pressure, raising the atoms to higher energy levels.

> **Expert tip**
>
> The continuous spectra produced by the cores of hot stars pass through their cooler outer layers, where the presence of elements results in an absorption spectra being observed on Earth. This is how astronomers are able to identify the elements present in distant stars.

Discrete energy and discrete energy levels

Revised ▢

- There is energy stored in the electric fields that exist between the charged particles of an atom (protons and electrons). But this energy can only exist in certain, *discrete* values. We use the term **discrete** to describe something which is individually separate, rather than continuous.
- Consider Figure 7.3. This is a simple representation of four energy levels within an atom. (In reality there will be many more than four levels.) $E_4 > E_3 > E_2 > E_1$, but other than that, Figure 7.3 does not represent any other information about the atom. A more detailed energy level diagram of hydrogen is shown later.
- When an electron in an atom moves between an energy level and a *lower* level, electromagnetic radiation energy is *emitted* of a certain precise frequency. The six possible transitions of Figure 7.3 are shown with blue arrows. Similarly, when an atom moves between one energy level and a *higher* level, electromagnetic radiation of a certain precise frequency has been *absorbed*.

> **Expert tips**
>
> Up to this point we have just referred to the energy levels of *atoms*, but more specifically, the transitions within the atoms that we are describing arise as a result of changes in the movement and position of electrons which, in classical physics, may be considered to move around the nucleus in fixed orbits. As such, the levels may also be commonly referred to as the energy levels of *electrons*.
>
> There are also important energy levels within the *nuclei* of atoms (discussed in Section 12.2 for HL students).
>
> The existence of discrete energy levels within atoms was discovered long before anyone knew *why* the levels existed. An understanding only became possible after the discovery of the wave properties of electrons (see Section 12.1 for HL students).

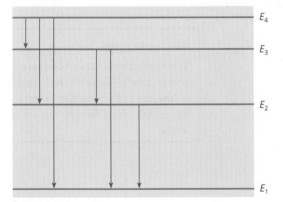

E_4

E_3

E_2

E_1

Figure 7.3

■ Photons

- When a single atom moves from a higher energy level to a lower energy level, a single, discrete amount of energy is emitted in the form of electromagnetic radiation (commonly light). This tiny 'bundle', or 'packet', of energy is called a **photon**. The frequency, f, of a photon is proportional to its energy, E: $E = hf$
- h is called **Planck's constant**. Its value is 6.63×10^{-34} J s.
- Remembering that $c = f\lambda$ (from Section 4.2), $E = hf$ can be re-written to directly show the relationship between the emitted wavelength and the photon's energy: $\lambda = \dfrac{hc}{E}$.
- Larger energy level transitions within an atom involve photons of higher frequencies (and shorter wavelengths).
- An atom can move from a lower energy level to a higher energy level if it *absorbs* a photon of the right energy.

> **Key concept**
>
> A **photon** is a quantum of electromagnetic radiation.
> The energy of a photon can be calculated from $E = hf$.

■ We describe electromagnetic radiation and energy levels as being **quantized**. This term may be used to describe anything that can only have certain discrete values. (Coins are an everyday example of a quantized system.) The minimum possible value of a quantized entity is called a **quantum** (plural: *quanta*). A photon is a quantum of electromagnetic radiation.

■ The fact that electromagnetic radiation behaves as a stream of very many tiny energy 'packets' called photons conflicts with the wave model discussed in Chapter 4. Some properties of radiation (interference, for example) require an explanation in terms of waves, but other properties (line spectra, for example) need the concept of photons for an explanation. Using both explanations is known as **wave–particle duality**.

■ The various properties of different parts of the electromagnetic spectrum (Section 4.2) can usually be explained by considering that the energy they transfer arrives in photons of different energies, rather than continuously.

> **Expert tip**
>
> Planck's constant plays a central role in atomic and nuclear (quantum) physics. In the IB physics course HL students will also meet Planck's constant when studying the wave nature of matter (de Broglie's hypothesis) and Heisenberg's uncertainty principle (Chapter 12).

QUESTIONS TO CHECK UNDERSTANDING

1   Calculate the energy carried by a photon emitted by a mobile phone if its frequency is 849 MHz.

2   a   The spectrum of helium contains a line of yellow light of wavelength 587 nm. How much energy is carried by a photon of this radiation?

    b   Suggest how the element helium was first discovered in the Sun from the spectrum detected on Earth.

3   A particular kind of photon carries an energy of 9.1×10^{-19} J.

    a   What is this energy when expressed in electronvolts?

    b   What is the wavelength of this photon?

    c   What part of the electromagnetic spectrum is it from?

4   A small radiant heater has a power of 800 W. It emits infrared and light with an average frequency of 5.0×10^{13} Hz. Estimate the number of photons emitted every second by this heater.

5   Compare the energy size of photons of light and photons of gamma rays, and use the difference to suggest why these two radiations have very different effects on the human body.

■ Transitions between energy levels

■ Figure 7.4 illustrates some of the many energy levels of hydrogen, labelled in both joules and electronvolts. Hydrogen, the simplest atom, has the simplest set of energy levels to understand and its study was of great historical importance.

> **Key concepts**
>
> When electrons move between energy levels, photons are emitted or absorbed. The photon energy (hf) equals the difference in energy levels.

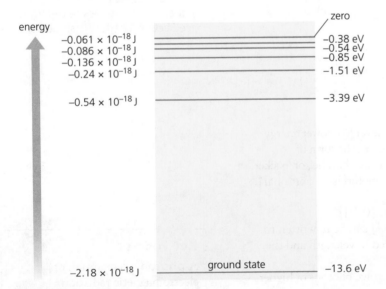

Figure 7.4

- If an electron in the ground state is given +13.6 eV of energy it will just be able to escape from a hydrogen atom, meaning that the atom would be *ionized*. The free electron is then said to have zero electrical potential energy, but when it was in the ground state its energy was 13.6 eV *less*, so in its ground state its energy is given a negative value (−13.6 eV). All the energy states of bound electrons have negative values for this reason.

Solving problems involving atomic spectra, including calculating the wavelength of photons emitted during atomic transitions

- We can use the difference in energies of any two levels to calculate the frequency (and/or wavelength) of a photon emitted or absorbed by a transition between those levels.
- For example, the first and third levels above the ground state in Figure 7.4 are −0.54 × 10⁻¹⁸ J and −0.136 × 10⁻¹⁸ J, so that the difference in energy is −0.404 × 10⁻¹⁸ J. $E = hf$ can be used to calculate the frequency of the photon which would be involved with this transition (6.09 × 10¹⁴ Hz, which is in the blue section of the visible spectrum)
- Alternatively, the wavelength can be calculated directly from $\lambda = \frac{hc}{E}$.

QUESTIONS TO CHECK UNDERSTANDING

6 Consider again Figure 7.4.

   a What frequency photon is emitted when an electron in an atom moves from the second energy level above the ground state down to the ground state?

   b Identify a possible transition that results in the emission of a photon of wavelength 4.4 × 10⁻⁷ m.

   c In which part of the electromagnetic spectrum is a photon which is emitted with the lowest frequency possible (when it moves between any of the levels shown)?

7 Figure 7.5 shows some of the energy levels of the atoms a simple gaseous element.

   a Calculate the frequency emitted by a transition from the first excited state.

   b What wavelength radiation will be absorbed if an atom is raised from its ground state to the third excited state?

   c What is the ionization energy of this element in joules?

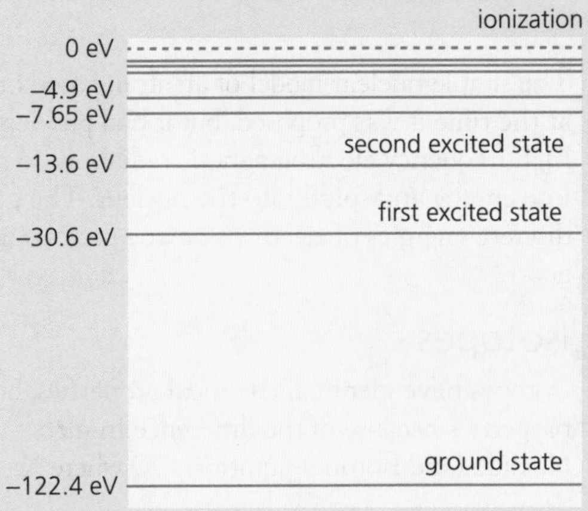

Figure 7.5

Atomic structure

Revised ☐

- A simple model of atomic structure involves a relatively small nucleus containing positively charged protons and neutral neutrons. Protons and neutrons are known as **nucleons** (particles from the nucleus). Negatively charged electrons move around (orbit) the nucleus because of the electric force between opposite charges.
- To an order of magnitude, a typical atom has a diameter of 10⁻¹⁰ m, while the nucleus at its centre is about 10⁻¹⁴ m in diameter.
- Figure 7.6 shows a common and historical visualization of an atom (in this example, nitrogen) which is, it should be noted, completely out of scale. However, this 100-year-old model, which is similar to planets moving around the Sun, useful as it is, should not be taken too literally. As discussed later in the chapter, scientists' ideas about atomic structure have been amended, but not in ways that are easy to represent in drawings.

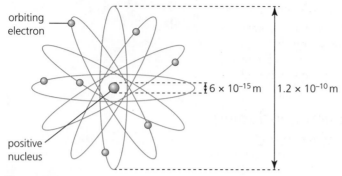

Figure 7.6

- The (rest) mass of the neutron, ($m_n = 1.675 \times 10^{-27}$ kg) is very nearly the same as the (rest) mass of the proton ($m_p = 1.673 \times 10^{-27}$ kg), but the electron's (rest) mass is much smaller ($m_e = 9.110 \times 10^{-31}$ kg). The charge on the proton is equal in magnitude to the charge on the electron ($\pm 1.60 \times 10\text{–}19\,C$).

> **Expert tip**
>
> The *rest mass* of any object or particle is its mass as measured by an observer who is at rest compared to the object (so that they have the same velocity). Einstein's theory of relativity (Option A) predicts that the mass of any object increases with its velocity relative to an observer, although the effect is only significant at speeds close to the speed of light.

- The simple nuclear model of an atom based on electrical forces was very useful at the time it was proposed, but it had problems: accelerating particles radiate high-frequency electromagnetic radiation, so that the electrons should rapidly lose energy and spiral into the nucleus. The model also does not explain the discrete energies of electrons or why protons in the nucleus are not repelled apart.

Isotopes

- Isotopes have identical chemical properties, but slightly different physical properties because of the difference in mass.
- Nuclide (and isotope) notation: $^A_Z X$, where X is the symbol of the element.

QUESTIONS TO CHECK UNDERSTANDING

8 $^{238}_{92}U$ is the most common isotope of uranium. How many protons, neutrons and electrons are there in an atom of this nuclide?

9 Oxygen has the proton number 8. It has two stable isotopes, one has 8 neutrons and the other 10.

 a Write down the two nuclide symbols for these isotopes.

 b Calculate, to 2 significant figures, the mass (kg) of a nucleus of the heavier isotope.

 c A sample of water, H_2O, contains both isotopes (in a ratio of about 400 : 1). Suggest why oxygen-18 molecules evaporate at a slower rate than oxygen-16 molecules.

10 Explain the difference between the terms 'nuclide' and 'isotope'.

> **Key concepts**
>
> The **proton number**, Z, is the number of protons in the nucleus of an atom. (Sometimes also called *atomic number*.) Atoms which have the same number of protons are atoms of the same element.
>
> The **nucleon number**, A, is the sum of the number of neutrons plus the number of protons. (Sometimes also called *mass number*.)
>
> The number of neutrons in the nucleus of an atom, the **neutron number**, N, is calculated from $A - Z$.

> **Key concepts**
>
> Two atoms which have the same number of protons, but different numbers of neutrons are known as **Isotopes** (of the same element).
>
> The term **nuclide** is used to describe a particular type of atom, as characterized by the contents of its nucleus.

Fundamental forces and their properties

Revised

- Coulomb's law (Section 5.1) can be used to determine the magnitude of the forces acting between protons in the nuclei of atoms. Typically, such forces are of the order of 100 N, which are enormous forces on such a small particles.

■ In order to explain how it is possible for positively charged protons to be packed close together in a nucleus without *strongly* repelling each other apart, we need to introduce the concept of a *strong nuclear force* attracting nucleons together. See Figure 7.7.

■ The strong nuclear force is considered to be one of the four **fundamental forces (interactions)** in the universe. More details of these forces are given in Section 7.3.

> **Key concept**
>
> The **strong nuclear force** acts between nucleons causing them to be packed closely together in the nucleus.

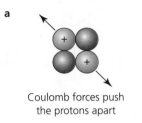
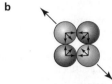

Coulomb forces push the protons apart

There must also be strong nuclear forces between the nucleons pulling them together

Figure 7.7

Nuclear stability

Revised ☐

■ The stability of any nucleus depends on the balance between the strong nuclear force, which attracts nucleons together, and the electric force repelling protons apart. This balance depends on the ratio of the number of neutrons to the number of protons. The pattern of stability can be represented on a *chart of the nuclides*, as shown in Figure 7.8.

■ The nuclei of nuclides with low proton numbers are stable if the number of neutrons is approximately equal to the number of protons, for example $^{12}_{6}C$. As the proton number increases, greater numbers of neutrons are needed if the nucleus is to be stable. For example, $^{116}_{50}Sn$ is a stable nuclide. Elements with a proton number of 83 or greater have no stable isotopes.

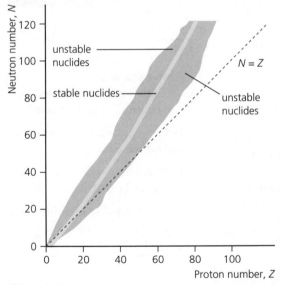

Figure 7.8

Radioactive decay

Revised ☐

■ An unstable nucleus emits *nuclear radiation* (alpha particles, beta particles or gamma rays) which carry energy away from the nucleus, thus making it more stable.

■ The rate of radioactive decay cannot be controlled because it is unaffected by chemical and physical properties, such as temperature.

■ The emitted radiation is often described as **ionizing radiation** because it causes atoms in the surrounding materials to lose electrons and form ions. This can cause chemical changes.

■ The cells of the human body may undergo chemical and biological changes as a result of exposure to ionizing radiation. This is dangerous to health.

> **Key concepts**
>
> **Radioactivity** is the name that we give to the emission of radiation by a nuclide.
>
> **Radioactive decay** is the process by which radioactive nuclides change into different elements after emitting particles. Decay is *spontaneous* and *random*.

■ Background radiation

- ■ We are all exposed to low levels of naturally occurring ionizing radiation in our everyday lives. This is called **background radiation** and it mostly originates from traces of radioactive materials in the ground and air, and cosmic rays from outer space.

■ Radioactivity experiments

- ■ Because radiation can be harmful to humans, there are a range of important safety precautions necessary when performing radioactivity experiments.
- ■ There are various types of *radiation detector* and knowledge of how they work is *not* needed for this course. (One common kind is called a *Geiger–Müller tube*.) Measurements are recorded as *counts*: numbers which indicate the amount of radiation detected in any given time; or as a **count rates**, for example the count per minute.
- ■ We cannot be sure that a counter detects all the radiation that is incident upon it. Its sensitivity may also vary for different radiations.
- ■ It may be necessary to subtract a *background count* from measurements made in radioactivity experiments in order to know the count that is due from a source alone.
- ■ Because of random variations in the emission of radiation from a source and from the 'background', it is normal for there to be variations in count rates taken under identical circumstances. This can be a significant problem affecting the accuracy of experiments with low count rates.

> **Key concept**
>
> The number of decays every second in a source is known as its **activity**. Activity is not easily measured, although we often assume that a *count rate* is proportional to a source's activity.

Alpha particles, beta particles and gamma rays

Revised ▢

- ■ There are three principle types of nuclear radiation: alpha particles, beta particles and gamma rays.
- ■ *Alpha particles* are given the symbol ^4_2He, or $^4_2\alpha$. They are doubly positively charged ($+2e$) and travel fast (about 10% of the speed of light, c). Their energy can be calculated using the equation for kinetic energy ($\frac{1}{2}mv^2$). All alpha particles emitted from the same nuclide have the same energy.
- ■ *Beta-negative particles* are singly negatively charged ($-e$) and given the symbol $^0_{-1}e$, or $^0_{-1}\beta$.
- ■ *Beta-positive particles* are given the symbol $^0_{+1}e$, or $^0_{+1}\beta$. A *positron* is identical to an electron, except that it is positively charged ($+e$). It is the *antiparticle* of an electron. There is more about antimatter in Section 7.3.
- ■ Beta particles can travel at speeds close to the speed of light. Their energy can be calculated using the equation for kinetic energy. Beta particles emitted from different atoms of the same nuclide have a range of different energies.
- ■ *Gamma rays* have no mass or charge. Their energy can be calculated using the equation for photon energy ($E = hf$).

> **Key concepts**
>
> **Alpha particles** are a combination of two protons and two neutrons (helium nuclei).
>
> **Beta-negative particles** are very high-speed electrons.
>
> **Beta-positive particles** are very high-speed **positrons**.
>
> **Gamma rays** are high-frequency electromagnetic radiation (photons).

QUESTIONS TO CHECK UNDERSTANDING

11 Explain what we mean when we say that radioactive decay is *spontaneous*.

12 Consider Figure 7.8. Predict the nucleon number of a stable isotope of ytterbium (proton number 70).

13 Calculate the approximate ratio of the mass of an alpha particle to the mass of a beta particle.

14 Calculate the energy (in MeV) of:

   a a beta particle travelling at a speed of $0.72c$

   b an alpha particle travelling at $0.05c$

   c a gamma ray of wavelength 1.0×10^{-12} m.

15 Describe two safety precautions necessary when working with radioactive materials.

16 Explain why the accuracy of radioactivity measurements is greater if the count rates are higher.

> **Expert tip**
>
> The fact that beta particles from the same source have a range of different speeds means that another (uncharged) particle must be emitted during beta decay (consider conservation of momentum). The other particle is a neutrino or an antineutrino, both of which are very difficult to detect.

Completing decay equations for alpha and beta decay

- After a transmutation, the resulting nuclide is often called a *daughter product*.
- The following balanced decay equations represent three possible transmutations:

  □ ^A_ZX → $^{A-4}_{Z-2}X$ + ^4_2He
  parent nuclide   daughter nuclide   alpha particle

  □ ^A_ZX → $^A_{Z+1}X$ + $^0_{-1}\beta$ + $\bar{\nu}_e$
  parent nuclide   daughter nuclide   beta-negative   antineutrino
                                        particle

  □ ^A_ZX → $^A_{Z-1}X$ + $^0_{+1}\beta$ + ν_e
  parent nuclide   daughter nuclide   beta-positive   neutrino
                                          particle

> **Key concept**
>
> When a particle is emitted during radioactive decay, a new element is formed. This process is called **transmutation**. Transmutations are represented in balanced **nuclear equations**.

- There is more about **neutrinos** and **antineutrinos** (antiparticles of each other) in Section 7.3. They are fundamental particles of *very* small mass and no charge.
- There are three different kinds of neutrinos/antineutrinos (see later). Those involved with beta decay are called *electron* neutrinos or antineutrinos.
- During gamma decay no new element is formed: there are no changes to the proton or nucleon number of the parent nuclide. Gamma rays are released from excited nuclei.
- A nucleus may undergo a series of radioactive decays, producing different elements by the emission of alpha and beta particles. This means that radioactive sources may contain a range of different nuclides. The relative proportions will depend on their *half-lives* (see later).

Absorption characteristics of decay particles

- Ionizing radiations will travel away from their source until they have transferred most or all of their energy in creating ions in the air or surrounding materials. We then say that they are **absorbed** and cannot penetrate any further. The **penetrating power** of radiation depends on its energy and how many ions it creates every centimetre.
- Gamma rays cause the least ions per centimetre, which means they penetrate furthest and are not easily absorbed. They can travel very large distances in air without significant absorption, although the radiation will lose *intensity* as it spreads out (approximating to an *inverse square law*). Gamma ray intensity is typically reduced by half by passing through about two or three centimetres of lead. See Figure 7.9.

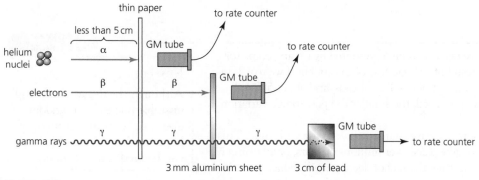

Figure 7.9

- Alpha particles create the most ions per centimetre and therefore only travel a few centimetres in air. See Figure 7.10, which shows the formation of ion pairs. Alpha particles cannot pass through thin paper.

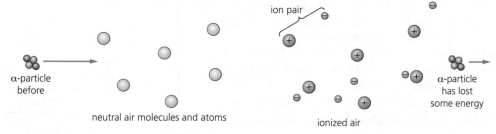

Figure 7.10

- Beta –negative particles are less ionizing than alpha particles and they typically travel about 1 m in air before they are absorbed. They are typically absorbed by about 2 or 3 mm of aluminium.
- The absorption of radiation can be used to identify the three different types of radiation.

Expert tip

The three types of radiation are also affected differently as they pass through electric and magnetic fields. Gamma rays are unaffected. Beta particles are deflected more than alpha particles because of their much smaller masses. When alpha or beta particles move perpendicularly across magnetic fields, the 'left-hand rule' can be used to predict the direction of the force that results in motion in the arc of a circle (see Section 5.4). Forces on alpha and beta particles in perpendicular electric fields produce parabolic paths.

QUESTIONS TO CHECK UNDERSTANDING

17 $^{14}_{6}$C is a radioactive nuclide which decays by beta-negative emission to nitrogen. Write an equation to represent this decay.

18 What is the decay equation for the decay of radium-222 by alpha emission to radon, Rn (proton number 86)?

19 The following transmutation occurs when a nuclide of protactinium decays: $^{230}_{91}$Pa → $^{230}_{90}$Th.

 a What kind of decay is this?

 b Write out the full decay equation for this transmutation.

20 Where on the chart of the nuclides (Figure 7.8) would you expect to find:

 a alpha particle emitters

 b beta-negative particle emitters?

21 Explain how you would use absorption tests to confirm that a source of radiation was only emitting beta particles.

22 An alpha particle of energy 3 MeV produced about 2×10^5 ion pairs in 4 cm of air before it was absorbed.

 a Estimate the average ionization energy of the molecules in the air.

 b If human skin is about 1000 times denser than air, estimate how far the alpha particles could travel into skin. State any one assumption you made.

Half-life

Revised

- The randomness of radioactive decay can be represented by tossing coins (or dice): we cannot predict the result of a single toss of a coin, but if we toss 1000 coins, we can be very sure that about 500 will be 'heads' and about 500 will be 'tails'. The more coins that are tossed, the smaller the percentage variation between reality and prediction.
- The count rate from any pure radioactive source will decrease with time. This decrease may be very quick, or take place over millions of years, or anything in between. But whatever the timescale, the *rate* of decrease will reduce over time as shown in Figure 7.11, which represents the count rate from any pure source. In theory, the count rate will never reduce to zero, although it will become vanishingly small.
- A radioactivity count rate reduces because, as time passes, there are fewer and fewer un-decayed nuclei remaining which are able to decay.
- This behaviour is commonly demonstrated with a dice *analogue* experiment, in which $\frac{1}{6}$ (chosen by one side of the dice) of an initially large number of dice are removed each time that all the remaining dice are thrown.
- The *half-life*, $T_{1/2}$, of a radioactive nuclide, is a way of quantifying the rate of a radioactive decay (or any other exponential decay). Half-life is also equal to the time it takes for the *activity* or the count rate to halve.
- If N is the initial number of radioactive nuclei of a pure sample, then after a time of one half-life, $\frac{1}{2}N$ will have decayed and $\frac{1}{2}N$ radioactive nuclei are left in

Key concept

Although the decay of any single unstable nucleus is a random event, the overall behaviour of large numbers of unstable nuclei can be predicted accurately.

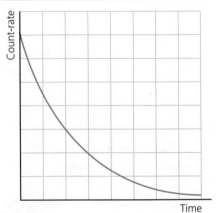

Figure 7.11

the sample. After two half-lives then there will be only $\frac{1}{4}N$ radioactive nuclei left in the sample and $\frac{3}{4}N$ will have decayed, etc.

Investigating half-life experimentally (or by simulation)

- If we wish to experimentally confirm the half-life of a nuclide it is necessary to use an isotope with a half-life of a suitable value – a few minutes would be ideal. The count rate should be measured over several half-lives, adjusted by deducting the background count, and a graph drawn of count rate against time. Figure 7.12 shows an example.

Determining the half-life of a nuclide from a decay curve

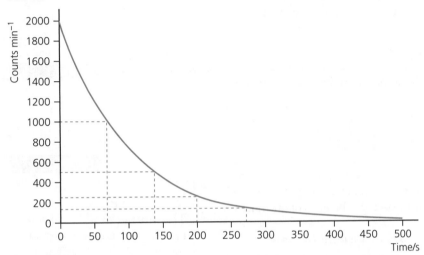

Figure 7.12

- The time for the count rate to reduce from *any* value to half of that value will equal the half-life of the nuclide. Accuracy can be improved by determining the average of several values taken from the same graph.
- The decay curve shown in Figure 7.12 has been drawn using experimental data (*count rates*) directly obtained from a radiation detector (adjusted for background count). A graph of how the number (or percentage) of un-decayed atoms, varied with time would produce the same half-life for the same nuclide, as would an activity–time graph, but in such cases, the data would not be so easily obtained *experimentally*.

> **Expert tip**
>
> This kind of curve seen in Figure 7.11 represents an *exponential decay*, in which a quantity always reduces to the same fraction (half, for example) in equal time intervals.

> **Key concept**
>
> **Half-life** is the time taken for half of any sample of unstable nuclei to decay.

> **Expert tip**
>
> Most radioactive sources are not pure, but contain a mixture of different nuclides. This is because, over time, there will have been a series of transmutations into a **decay series (chain)** of different elements. As a result, such sources will not have a single, simple half-life. To obtain a pure source in a school laboratory it will be necessary to extract a suitable element from a mixture. This can be done by using a selective solvent in a sealed container.

> **Common mistake**
>
> The count rate measured by a radiation detector is often confused with the *activity* of the source emitting the radiation. They are not the same, but can usually be assumed to be proportional to each other.

QUESTIONS TO CHECK UNDERSTANDING

(Note that, in this chapter, calculations will only involve *whole numbers* of half-lives.)

23 a What is the half-life of the nuclide represented in Figure 7.12?

  b Predict the count rate after 475 s.

24 A pure radioisotope has a half-life of 2 minutes.

  a What fraction of the atoms remains undecayed after 6 minutes?

  b What percentage has decayed after 8 minutes?

25 A student measures the count rate from a pure radioactive source every thirty seconds and records the following results (s⁻¹): 18.4, 12.3, 8.9, 6.2, 4.1, 3.4, 2.6. The background count at the time of the experiment was 0.4 s⁻¹. Draw a graph of these results and use it to determine the half-life of the isotope.

26 ^{60}Co has a half-life of 5.3 years. If a source of this nuclide is no longer useful after its activity has fallen below 10% of its initial value, *estimate* the number of years before it has to be replaced.

27 ^{99}Tc is a very widely used isotope in hospitals, where it is injected into patients to 'trace' its movement in the body (by detecting radiation that radiates out of the body). This radionuclide emits gamma rays and has a half-life of 6 hours.

  a Why is it important that its half-life is a matter of hours?

  b Why is a nuclide which emits gamma rays used?

  c What percentage remains in the body after one day?

28 Describe an experiment to determine the half-life of a nuclide.

Expert tip

In Chapter 12, for HL students, we learn how to use the mathematics of exponentials to perform calculations involving *any* time period, not just those involving whole numbers of half-lives.

NATURE OF SCIENCE

■ Accidental discovery

The unintentional discovery of radioactivity by Henri Bequerel is a well-known story in science, but there have been many notable 'accidental' discoveries. This, in itself, is not surprising because events rarely go according to expectations. However, it is in the *response* to unexpected, or seemingly insignificant, discoveries that we should recognize the true talents of the great scientists.

7.2 Nuclear reactions

Revised ☐

Essential idea: Energy can be released in nuclear decays and reactions as a result of the relationship between mass and energy.

The unified atomic mass unit

Revised ☐

- The SI unit mass, the kilogram, is inconveniently large when working with the much, much smaller masses of atomic particles. A new unit of mass needs to be defined for use in atomic physics. It is based on the mass of one nucleon, but needs to be defined very precisely.
- We have seen that protons and neutrons do not have identical masses and that their mass also varies with relative velocity. We will also see later in this section that the mass of nucleons varies slightly when they are combined in different ways.
- The *unified atomic mass unit*, u, is widely used for recording the masses of atomic particles.

> **Key concept**
>
> The **unified atomic mass unit**, u, is defined as one-twelfth of the mass of an isolated carbon-12 atom. $1u = 1.661 \times 10^{-27}$ kg

Equivalence of mass and energy

Revised ☐

- Following his work on relativity (Option A), Einstein was the first to state that mass and energy are the same property of a physical system. This is described as the **equivalence of mass and energy**. It is considered to be misleading to state that matter is *converted* into energy or vice versa.
- The increase in mass, Δm, that accompanies an increase in energy, ΔE, can be calculated from $\Delta E = \Delta mc^2$, where c is the speed of light. Since $c^2 \approx 10^{17}$, it should be clear that in everyday life, such increases in mass ($= \Delta E/c^2$) are immeasurably small.
- In atomic physics (because of the sizes of the forces within nuclei) relatively large amounts of energy are transferred from very small masses, so that, unlike the macroscopic world, changes of masses can become significant and may be measurable. This means that in atomic physics, mass has to be included in energy calculations, and we often refer to the *conservation of mass–energy* rather than the conservation of just energy (or mass).
- We now have an alternative unit for atomic masses: $1u = (1.661 \times 10^{-27}) \times (2.998 \times 10^8)^2 = 1.493 \times 10^{-10}$ J. This can be converted to the usual energy units for atomic physics (MeV) by dividing by $(1.602 \times 10^{-19} \times 10^6)$, to give $1u = 931.5$ MeVc^{-2}. A mass of 1u (1.661×10^{-27} kg) is equivalent to an energy of 931.5 MeV.
- The rest mass of the proton was given in Section 7.1 as $m_p = 1.673 \times 10^{-27}$ kg; this is equal to **1.007276 u** or **938 MeVc^{-2}**.
- The rest mass of the neutron was given in Section 7.1 as $m_n = 1.675 \times 10^{-27}$ kg; this is equal to **1.008665 u** or **940 MeVc^{-2}**.
- The rest mass of the electron was given in Section 7.1 as $m_e = 9.110 \times 10^{-31}$ kg; this is equal to **0.000549 u** or **0.511 MeVc^{-2}**.

> **Key concepts**
>
> Mass and energy are two ways of measuring the same thing (mass-energy). If the energy of a system is increased in any way (for example by heating it, or making it move faster, or moving it away from the Earth), then there must be an equivalent increase in mass.
>
> Nuclear physics involves relatively large amounts of energy and small masses, so that the changes of mass which occur when energy is transferred become significant. They can be calculated from $\Delta E = \Delta mc^2$.

> **Common mistakes**
>
> The equation $\Delta E = \Delta mc^2$ is perhaps more commonly seen in the form $E = mc^2$. However, we are using this form because it stresses *changes* to E and m. Using the equation $E = mc^2$ may suggest only the *total* conversion of a mass to energy.

29 a   What is the total mass (in u) of: **i** six separate protons, **ii** six separate neutrons, **iii** six separate electrons?

b   Add your answers together to give the total mass of the separate particles of a $^{12}_{6}$C atom.

c   Compare your answer to part **b** to the defined mass of the $^{12}_{6}$C atom. (The reason for the difference is given below.)

30 The mass of an atom of oxygen-16 is 15.999 u. Express this mass in:

a   kg

b   MeV.

31 A 1500 kg car is accelerated from rest to 20 m s⁻¹. Calculate its kinetic energy and the equivalent increase in mass.

Mass defect and nuclear binding energy

Revised ☐

■ In order to separate all the nucleons in a nucleus it would be necessary to *supply* energy to overcome the strong nuclear forces holding them together. The same amount of energy would be *released* if a new nucleus was formed from the same (separated) nucleons. These are only 'thought experiments' because it is not practically possible. The energy that would be involved is called the *binding energy* of the nucleus.

> **Key concepts**
>
> **Nuclear binding energy** is the amount of energy needed to completely separate *all* of the nucleons of a nucleus.
>
> Nuclear binding energy has a mass equivalent which is called the **mass defect** of the nucleus.

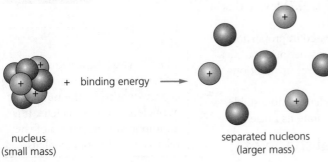

nucleus
(small mass)

separated nucleons
(larger mass)

Figure 7.13

■ Alternatively, *nuclear binding energy* can be considered to be the energy that would be released if a nucleus was formed from separate nucleons. See Figure 7.13.

■ Because of the energy difference, the total mass of all the nucleons together in a nucleus is less than the sum of the masses of its nucleons when they are separated. This difference is called a *mass defect*.

■ Solving problems involving mass defect and binding energy

32 a   What is the mass defect of a $^{12}_{6}$C atom (see question 29)?

b   What is the binding energy of a $^{12}_{6}$C nucleus (in MeV)?

33 Iron-56 has a mass defect of 0.528479 u.

a   What is this mass expressed in kilograms?

b   What is the binding energy of this nucleus (in MeV)?

34 An atom of $^{16}_{8}$O has a mass of 15.9994 u. What is its mass defect (in u)?

Solving problems involving the energy released in radioactive decay

- When an unstable nucleus emits energy (the kinetic energy of a particle or the electromagnetic energy of a photon), in order for the *mass–energy* of the system to remain constant, there must be a slight decrease in the total mass of all the particles involved.

QUESTIONS TO CHECK UNDERSTANDING

35 The nuclide $^{24}_{11}$Na decays by beta-negative emission to magnesium.

  a Write an equation for this decay.

  b If the rest mass of the sodium atom was 23.99096 u and the rest mass of the magnesium atom was 23.98504 u, determine the energy (MeV) of the emitted beta particle.

36 An atom of $^{226}_{88}$Ra has a mass of 226.02541 u and emits an alpha particle of kinetic energy 4.77 MeV. The resulting nuclide is an isotope of radon (Rn).

  a Write an equation for this decay.

  b Determine the mass (u) of the radon atom. (Mass of an alpha particle = 4.00151 u.)

Average binding energy per nucleon

Revised

- For example, the binding energy of the lithium-7 nucleus is 37.7 MeV, so that its binding energy per nucleon is 5.4 MeV.
- We might expect that the binding energy of nuclei increased in proportion to the number of nucleons, so that the average binding energy per nucleon was approximately the same, but there are significant and interesting variations. See Figure 7.14.
- A higher binding energy per nucleon means that the nucleus is more stable, because more energy per nucleon would be needed to separate its nucleons.

> **Key concept**
>
> The **average binding energy per nucleon** is the total binding energy divided by the number of nucleons in the nucleus. It is an important indication of the relative stability of a nucleus.

Sketching and interpreting the general shape of the curve of average binding energy per nucleon against nucleon number

- Average binding energy per nucleon rises with nucleon number to a maximum (at nickel-62: 8.8 MeV) and then gradually decreases with increasing nucleon number.

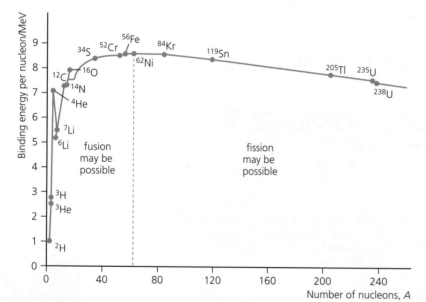

Figure 7.14

> **Expert tip**
>
> We have represented binding energy as positive, but if we consider that separated nucleons have zero potential energy, then binding energies should be quoted with negative values to represent the fact that if enough energy is added to the nucleus to separate the nucleons, the total energy will then be zero.

37 What is the average binding energy per nucleon of a $^{16}_{8}$O nucleus (use data from question 34)?

38 Use Figure 7.14 to:

  a determine the binding energy of a $^{184}_{74}$W nucleus

  b estimate the average binding energy per nucleon of $^{132}_{56}$Ba.

Nuclear fission and nuclear fusion

Revised ☐

- As can be seen in Figure 7.15a, *nuclear fission* is only possible for *massive* nuclei because *if* fission can be induced, the average binding energy per nucleon increases in the process. Fission will result in greater nuclear stability and the release of energy (in the form of kinetic energy of the newly formed nuclei and neutrons, and the electromagnetic energy of photons: see below).
- As a result of the energy released during fission, the mass of the nuclei must have decreased, although the average binding energies per nucleon have increased (more energy is required to separate the nucleons than before).
- In practice, nuclear fission is only possible with a few nuclides. The most well-known example is uranium-235, used in nuclear power stations.
- Uranium-235 can be induced to undergo nuclear fission when it is bombarded with slow-moving neutrons. For example,

$$^{1}_{0}n + ^{235}_{92}U \rightarrow ^{236}_{92}U \rightarrow ^{141}_{56}Ba + ^{92}_{36}Kr + 3^{1}_{0}n + \text{photons}$$

- When nuclear fission occurs continuously, very large quantities of energy can be released and this can be used in nuclear reactors (see Section 8.1) and nuclear weapons.
- *Nuclear fusion* is only possible with light nuclei because only those nuclei can be combined to produce a nucleus which has a larger average binding energy per nucleon. See Figure 7.15b. Because energy is released, the mass of the system will be less than the initial masses of the smaller nuclei.
- The following is an example of nuclear fusion, in which the nuclei of two isotopes of hydrogen combine: $^{2}_{1}H + ^{3}_{1}H \rightarrow ^{4}_{2}He + ^{1}_{0}n + 17.6\,\text{MeV}$ of energy.
- Fusion reactions occur inside stars and are responsible for generating new elements from hydrogen and helium.
- Nuclear fusion can occur in stars because the temperature and the particle density are extremely high. Although nuclear fusion has been achieved in experimental reactors for very short periods of time on Earth, sustained release of energy has not yet been possible.

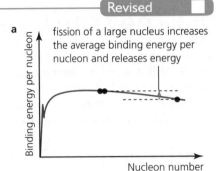
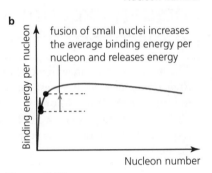

Figure 7.15

> **Key concepts**
>
> **Nuclear fission** is the process in which a massive nucleus splits into two smaller, more stable nuclei.
>
> **Nuclear fusion** occurs when two small nuclei join together to make a larger, more stable nucleus.
>
> Both processes increase average binding energy per nucleon and release significant amounts of energy, so that the total mass is reduced.

■ Solving problems involving the energy released in nuclear fission and nuclear fusion

39 Figure 7.16 represents an alternative fission of $^{235}_{92}$U.

  a Write a balanced nuclear equation for this reaction. (The proton number of rubidium is 37.)

  b Determine the energy released in this reaction. (Mass of ^{90}Rb = 89.915, mass of ^{143}Cs = 142.927, mass of $^{235}_{92}$U = 235.044.)

40 $^{2}_{1}$H (deuterium) can fuse with $^{3}_{1}$H (tritium) to form helium-4 and a neutron. Confirm that about 18 MeV of energy is released in this process. (Mass of $^{2}_{1}$H = 2.0136 u, mass of $^{3}_{1}$H = 3.0160 u, mass of $^{4}_{2}$He = 4.0020 u.)

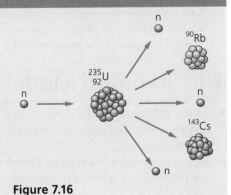

Figure 7.16

■ Patterns, trends and discrepancies

There are about 340 naturally occurring nuclides on Earth, of which about 85 are unstable. An enormous amount of physical data is available on these nuclides, the analysis of which requires the identification of patterns and trends in their behaviour. Figures 7.8 and 7.14 show two common ways of representing some of this data. Once patterns have been identified, scientists can use that information to make predictions about other nuclides and isotopes.

7.3 The structure of matter

Revised ☐

Essential idea: It is believed that all the matter around us is made up of fundamental particles called quarks and leptons. It is known that matter has a hierarchical structure with quarks making up nucleons, nucleons making up nuclei, nuclei and electrons making up atoms, and atoms making up molecules. In this hierarchical structure, the smallest scale is seen for quarks and leptons (10^{-18} m).

Describing the Rutherford-Geiger-Marsden experiment that led to the discovery of the nucleus

Revised ☐

- The model of an atom with a central nucleus was first proposed by Ernest *Rutherford* following an investigation of alpha particle scattering by thin gold foil (the **Rutherford-Geiger-Marsden experiment**, see Figure 7.17).
- Only a few alpha particles were scattered through large angles, which suggested that the nucleus is very much smaller than the whole atom. See Figure 7.18.

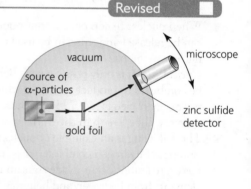

Figure 7.17

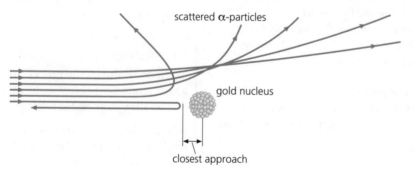

Figure 7.18

- By using alpha particles of different energies and/or targets of different metals, Rutherford was able to confirm that the scattering closely followed Coulomb's inverse square law (of electric repulsion).
- In the hundred years following Rutherford's (Figure 7.20) work a large number of *sub-atomic particles* have been discovered (most are unstable and have only been created in very high-energy physics laboratories). The structure of matter has been the subject of an enormous amount of research in many countries. The latest model of the particles in the universe is called the *Standard Model*.

The Standard Model

Revised ☐

- In the **Standard Model** there are only four types of *elementary* particle: *leptons*, *quarks* and *exchange particles* (also called *gauge bosons* and the *Higgs boson*).
- Figure 7.19 summarizes the classification of both types of particle. More details are provided in the rest of this section.

> **Key concept**
>
> The scattering pattern of positively charged alpha particles by thin gold foil provides evidence that they are repelled by small positive nuclei at the centres of atoms.

> **Expert tip**
>
> In Chapter 12 (HL) we see how equating the kinetic energy of the alpha particles to the electric potential energy between them and the nucleus can lead to an estimate of nuclear radius.

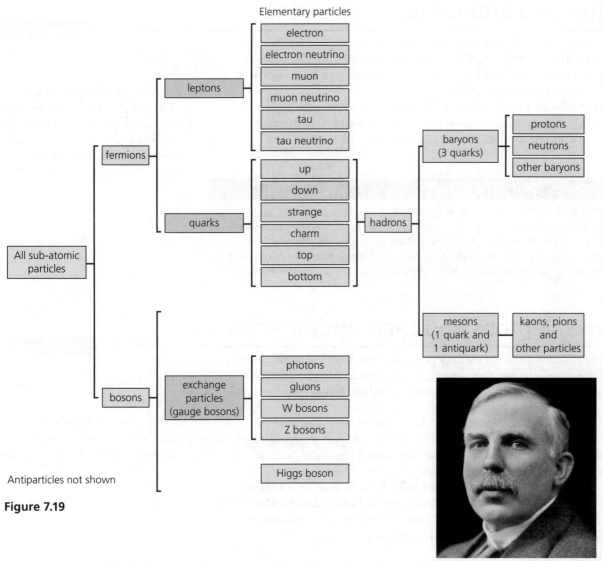

Elementary particles

Figure 7.19

Antiparticles not shown

Figure 7.20 Ernest Rutherford

■ Figure 7.21 is a chart of the Standard Model of all the elementary particles. Neither Figure 7.19 nor Figure 7.21 includes antiparticles.

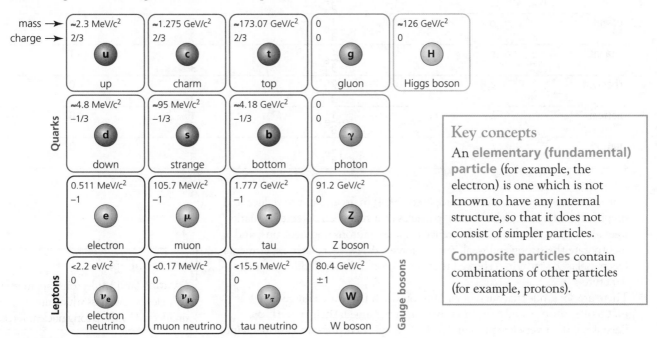

Figure 7.21

> **Key concepts**
>
> An **elementary (fundamental) particle** (for example, the electron) is one which is not known to have any internal structure, so that it does not consist of simpler particles.
>
> **Composite particles** contain combinations of other particles (for example, protons).

Matter and antimatter

Revised ☐

- We saw in Section 7.1 that the positron is the antiparticle of the electron.
- An antiparticle (**antimatter**) can only exist for a short time because when it meets a particle they will both **annihilate** and release a large amount of energy in the form of gamma rays.
- Antiparticles are usually represented by using a bar above the symbol, for example $\bar{p}$ represents an antiproton (e^+ is widely used for the positron).
- A few uncharged particles (for example, the photon) are their own antiparticles.

> **Key concept**
>
> For every particle there is an **antiparticle** which has the same mass as the particle, but with the opposite charge (if charged) and opposite quantum numbers.

> **Expert tip**
>
> Antimatter exists briefly on Earth and elsewhere as a consequence of beta particle decay (or pair production: see Section 12.1 for HL), or in high-energy physics laboratories. Scientists are still uncertain of the reason why the universe contains so much more matter than antimatter. This is one of the important unanswered questions in astrophysics concerning the early development of the universe.

Quarks, leptons and their antiparticles

Revised ☐

- Quarks and antiquarks are the only particles which have fractional charge (either $+\frac{2}{3}e$ or $-\frac{1}{3}e$), but they cannot exist as separate particles (charges). Quarks are *always* joined in combinations, such that, if they have an overall charge, it is $\pm e$, not a fraction of e.
- There are six kinds (**flavours**) of quark: **up**, **down**, **strange**, **charm**, **bottom**, **top**. The up and down quarks are very common in the universe; the others can only be created on Earth in high-energy physics laboratories, or when cosmic rays strike the atmosphere.
- The basic properties of quarks are shown in Table 7.1. Antiquarks are not included. **This table is similar to the one provided in the IB** *Physics data booklet*. (*Baryon number* and *strangeness* are explained later.) There is some uncertainty about the masses of individual quarks because they cannot be isolated.

> **Key concept**
>
> **Quarks** are the elementary particles from which protons, neutrons and mesons are composed.

Table 7.1

Flavour	Symbol	Electric charge	Baryon number	Strangeness
up	u	$+\frac{2}{3}e$	$\frac{1}{3}$	0
down	d	$-\frac{1}{3}e$	$\frac{1}{3}$	0
strange	s	$-\frac{1}{3}e$	$\frac{1}{3}$	−1
charmed	c	$+\frac{2}{3}e$	$\frac{1}{3}$	0
bottom	b	$-\frac{1}{3}e$	$\frac{1}{3}$	0
top	t	$+\frac{2}{3}e$	$\frac{1}{3}$	0

- The existence of quarks was first proposed (1964) in an attempt to reduce and simplify the number of 'elementary' particles that had been detected. Quarks were also used to explain the difference between protons and neutrons, and incorporate the strong and weak nuclear forces. A few years later experiments (deep inelastic scattering) showed that the proton did indeed have an inner structure.
- There are six kinds (flavours) of leptons, of which three are charged ($-e$): **electrons**, **muons** and **taus**, and three are uncharged: their **neutrinos**. Because of their very low mass and lack of charge, neutrinos and antineutrinos are very difficult to detect.

> **Key concept**
>
> **Leptons** (and antileptons) are mostly particles of low mass which are unaffected by the strong nuclear force.

- Electrons are stable and by far the most common charged lepton in the universe. Muons and taus are unstable and on Earth can only be produced in high-energy physics laboratories.
- The basic properties of leptons are shown in Table 7.2. (Antileptons are not included.) **This table is similar to the one provided in the IB *Physics data booklet*.** (*Lepton number* is explained on page 126.)

Table 7.2

Lepton	Symbol	Electric charge/e	Rest mass (MeV c^{-2})	Lepton number
electron	e	−1	0.511	1
electron neutrino	ν_e	0	very small, not known	1
muon	μ	−1	106	1
muon neutrino	ν_μ	0	very small, not known	1
tau	τ	−1	1780	1
tau neutrino	ν_τ	0	very small, not known	1

QUESTIONS TO CHECK UNDERSTANDING

41 a Write a brief account of the experiment shown in Figure 7.17.

  b Explain how the results of this experiment led to the nuclear model of the atom.

42 a Sketch the path of an alpha particle deflected by about 90° when it passes near a $^{197}_{79}$Au nucleus.

  b Add a second line to show the approximate path of an alpha particle which approached on the same path but with twice the energy of the first.

  c Explain how the angles of deflection would change if the alpha particles were fired at silver atoms (proton number 47).

43 Use Coulomb's law to calculate the forces acting between an alpha particle and a $^{206}_{82}$Pb nucleus when they are 3.0×10^{-10} m apart.

44 Explain why a neutron is not considered to be an elementary (fundamental) particle.

45 Write down the symbol and charge of a charm antiquark.

46 Lepton and quarks are collectively called fermions. Including antiparticles, how many types of elementary fermions are there in the universe?

Hadrons, baryons and mesons

Revised ▢

- Hadrons can be classified into two kinds: *baryons* (3 quarks) and *mesons* (2 quarks: a quark and an antiquark). Other combinations (for example 4 quarks) may be possible.
- Examples are shown in Figure 7.22. Protons and neutrons are the most common kinds of baryon, but there are others, like the various kinds of lambda particles.

> **Key concepts**
>
> Composite particles containing quarks are called **hadrons**.
>
> **Baryons** all contain three quarks.
>
> **Mesons** all contain one quark and one antiquark.

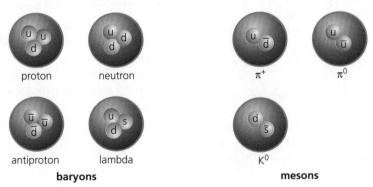

baryons   **mesons**

proton   neutron π^+ π^0

antiproton   lambda K^0

Figure 7.22

Describing protons and neutrons in terms of quarks

■ All nucleons and their antiparticles contain three quarks. An *antiproton*, for example, contains two up antiquarks and one down antiquark ($\bar{u}\bar{u}\bar{d}$)

Describing mesons in terms of quarks

■ There are many types of meson (such as pions and kaons) and antimesons. They are all unstable particles. There is no requirement in the IB Physics course to remember the names or compositions of mesons.

■ The concept of **strangeness** was first introduced when some mesons were found to have 'strange' properties. This unusual behaviour was then explained by proposing that the meson contained a different kind of quark: a *strange quark*, given a strangeness of −1.

> **Key concepts**
>
> **Protons** are composite particles which contain two up quarks and one down quark (uud).
>
> **Neutrons** contain one up and two down quarks (udd).

Confinement

Revised ■

■ Quarks can only exist in combination with other quarks or antiquarks.

Describing why free quarks are not observed

■ *Quark confinement* occurs because the very large force between quarks does not decrease with greater distance between them. If an increasingly large amount of energy was supplied in an attempt to separate bound quarks, it would only result in the formation of a new quark–antiquark pair (which then combine to form a meson) before the original quarks could be separated as free particles. In other words, adding energy creates more quarks, rather than separates quarks.

> **Key concept**
>
> **Quark confinement** refers to the fact that quarks do not exist as free individual particles, but only bound together (as hadrons).

QUESTIONS TO CHECK UNDERSTANDING

47 Write down the composition of two possible mesons and their antiparticles. Include at least one meson which contains a strange quark.

48 What are the composition, symbol and properties of an anti-neutron?

49 The Σ⁺ particle contains two up quarks and one strange quark.

   a What category of particle is this?

   b What is its charge?

   b It has a rest mass of 1189 MeVc^{-2}. What is this mass expressed in atomic mass units (u)?

The conservation laws of charge, baryon number, lepton number and strangeness

Revised ■

■ In any sub-atomic particle decay, or interaction between sub-atomic particles, certain properties must always be conserved. These properties are quantized and represented by **quantum numbers**.

■ The **charge number** of any particle will be 0 or a multiple of $\pm\frac{1}{3}$. For example, the charge on a proton is +1 ($\frac{2}{3} + \frac{2}{3} - \frac{1}{3}$), the charge of an up antiquark is $-\frac{2}{3}$, and the charge of the K⁰ meson shown in Figure 7.22 is zero ($-\frac{1}{3} + \frac{1}{3}$)

■ The **baryon number** of a particle is $\frac{1}{3}$ of (the number of quarks less the number of antiquarks). A single quark has a baryon number of $\frac{1}{3}$. A proton or neutron has a baryon number of 1. All mesons have baryon numbers of zero.

■ The **lepton number** of a particle is the number of leptons less the number of antileptons. All leptons have a lepton number of +1 and all antileptons have a lepton number of −1.

■ A **strangeness number** (−1) is only given to those hadrons which contain a strange quark. A hadron containing an antistrange quark is given the strangeness number +1.

> **Key concept**
>
> *Charge number, baryon number* and *lepton number* are *always* conserved in any interaction. The *strangeness number* is conserved in strong and electromagnetic interactions.

Applying conservation laws in particle reactions

- The quantum conservation rules help us predict what can, and what cannot, happen in particle reactions. (There are also other important quantum numbers which are *not* included in the IB Physics course.)
- Beta particle decay (Section 7.1) provides a suitable example: We might suppose that a neutron within an unstable nucleus decays to an electron and a proton ($n \rightarrow p^+ + e^-$), but lepton number is not conserved, so this is not possible. The addition of an anti-lepton (an electron antineutrino) to the right-hand side of the equation then represents the correct reaction.

QUESTIONS TO CHECK UNDERSTANDING

50 Write down equations which represent the changes within a nucleus which is undergoing:

  a beta-negative decay

  b beta-positive decay.

51 A positive pi meson (pion) consists of an up quark and a down antiquark. It is proposed that it could decay into an antimuon (positive muon) and a muon neutrino.

  a Write an equation for this possible decay.

  b Use the conservation laws to determine if it is possible.

52 Might it be possible for a tau particle to decay into two hadrons and a tau antineutrino? Explain your answer.

The nature and range of the strong nuclear force, weak nuclear force and electromagnetic force

Revised

- The nature of the four fundamental forces (interactions) was outlined in Section 7.1. This is summarized in Table 7.3. Forces acting 'at a distance' across empty space require some kind of explanation. This is achieved using the concept of *exchange particles*.

Comparing the interaction strengths of the fundamental forces, including gravity

Table 7.3

Fundamental interaction	Acts on	Approximate relative size	Range/m	Exchange particle
strong nuclear	quarks and gluons	1	10^{-15}	gluon (g)
electromagnetic	all charged particles	10^{-2}	infinite, but reduces with an inverse square law	photon (γ)
weak nuclear	quarks and leptons	10^{-6}	10^{-18}	W or Z bosons
gravitational	all masses	10^{-38} (not significant for individual particles)	infinite, but reduces with an inverse square law	graviton (to be confirmed)

- Some of the information shown in Table 7.3 is contained in the data booklet (but not the relative sizes or ranges of the forces).

Exchange particles

Revised

- Exchange particles are described as **virtual particles**, because they exist for only unmeasurably short periods of time and so cannot be detected. The **photon** is the exchange particle for the electromagnetic interaction, the **gluon** for the nuclear strong force, and **W** and **Z bosons** for the weak nuclear force. The **graviton** has been proposed as the exchange particle for gravitational forces (but not confirmed).

■ All gauge bosons are fundamental (elementary) particles of the universe.

■ Describing the mediation of the fundamental forces through exchange particles

■ When we observe a change of momentum we assume that an unbalanced force has acted (Section 2.2), even if we cannot see what caused the force. (Imagine, for example, watching a windmill starting to turn.) Fundamental forces acting across space are explained by the exchange of particles between the objects experiencing the forces. Each time a particle is emitted or received by an object, a force is exerted on that object.

■ The name 'virtual' and the fact that they cannot be detected should not suggest that the exchange particles are not real (imaginary).

■ Figure 7.23 shows a widely used broad analogy: each time the ball is thrown or received the person experiences a force and moves backwards. This model represents a repulsive force, but modelling attractive forces is less easy (some sources use a boomerang analogy).

■ A full explanation of exchange particles is not required in this course, and needs an appreciation of the *Heisenberg uncertainty principle* (Chapter 12, for HL students).

■ The larger an exchange particle, the shorter the range of the force that it *mediates* (carries between particles). Since photons (and gravitons) have zero rest mass, their range is infinite. At the other extreme, W and Z bosons have significant masses, which is the reason that they can only act over very short distances. Gluons have low mass.

Feynman diagrams

Revised

■ The exchange of virtual particles is best represented on *Feynman diagrams*. These diagrams attempt to give a visual representation of (unobservable) particle interactions. However, it is important to realize that Feynman diagrams represent interactions, *not* particle trajectories. That is, the precise direction of the lines is not important.

■ Sketching and interpreting simple Feynman diagrams

■ Essential features:
  □ Each interaction is represented by a point (*vertex*).
  □ The time before the interaction is to the left of the vertex, the time after is to the right of the vertex (but some sources use down/up).
  □ Observable particles are shown with straight lines to the right, with arrows. For antiparticles the arrow is reversed. There is always one arrow pointing to a vertex and one arrow pointing away.
  □ The direction of the line representing a particle changes after the interaction (to show that it has been affected).
  □ Exchange particles are represented by wavy lines, without an arrow.
  □ Each vertex joins two straight lines and a wavy line.
  □ The conservation laws can be applied at every vertex.

■ Students are not expected to remember particular Feynman diagrams. Some examples are shown in Figures 7.24–7.26.
  □ Figure 7.24 shows two vertices representing the mutual repulsion of two positrons, involving the exchange of a virtual photon.
  □ Figure 7.25 represents beta-negative decay and the importance played in that process by the W⁻ boson.

Figure 7.23

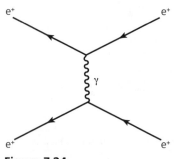

Figure 7.24

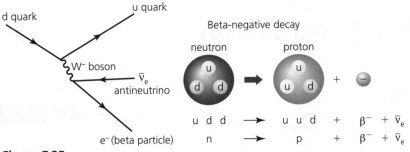

Figure 7.25

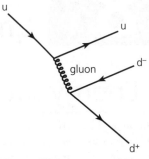

Figure 7.26

☐ Figure 7.26 shows a gluon involved in the formation of a quark–antiquark pair.

The Higgs boson

- The **Higgs field** is believed to occupy everywhere in the universe with a constant strength.
- When elementary particles interact with the Higgs field they acquire mass. The **Higgs boson** is the key boson in this interaction. It is a fundamental particle with a large mass.
- The discovery of the Higgs boson in 2012 at CERN was much anticipated and finally confirmed the process by which elementary particles acquire their mass.

> **Key concept**
>
> All particles gain the property that we call 'mass' by interacting with the Higgs field which is thought to be constant throughout the universe. The Higgs boson is the particle responsible for this interaction.

QUESTIONS TO CHECK UNDERSTANDING

53 Use Coulomb's law and Newton's law of gravitation to compare the sizes of the electro (magnetic) force and gravitational forces acting between a proton and an electron separated by 0.1 nm.

54 List the fundamental forces which can act on

 a   electrons

 b   quarks

 c   photons.

55 Describe the interaction represented in Figure 7.27.

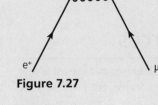

Figure 7.27

56 Draw a Feynman diagram to represent an electron and a positron annihilating to produce a neutrino pair.

NATURE OF SCIENCE

■ Predictions

Many discoveries have been made in particle physics because physicists have designed experiments to look for something that they had already predicted (the discovery of the Higgs boson being a recent example). In particle physics the belief in 'simplicity' and the symmetry of ideas is a particularly powerful driving force that promotes investigation of any gaps or inconsistencies in current patterns of knowledge.

■ Collaboration

The benefits of mutual collaboration between scientists is probably self-evident and modern research is very much characterized by teamwork. The very considerable expenses of research into nuclear physics make international collaboration vital in order to avoid unnecessary duplication of efforts and resources.

8 Energy production

8.1 Energy sources

Revised ☐

Essential idea: The constant need for new energy sources implies decisions that may have a serious effect on the environment. The finite quantity of fossil fuels and their implication in global warming has led to the development of alternative sources of energy. This continues to be an area of rapidly changing technological innovation.

Primary energy sources

Revised ☐

- Solar radiation, uranium and crude oil are examples of primary energy sources.
- Energy sources that use chemical or nuclear reactions to provide thermal energy or mechanical power are commonly called **fuels**.

> **Key concept**
> A **primary energy source** is a source which occurs naturally and has not been converted in any way.

Electricity as a secondary and versatile form of energy

Revised ☐

- *Electricity* is an example of a *secondary energy source*.
- The usefulness and *versatility* of electricity cannot be underestimated. Not only can electricity be converted to or from most other forms of energy relatively easily, it can be transferred around the country by cables with minimal energy dissipation.

> **Key concept**
> A **secondary energy source** is one that has been converted from a primary source.

Renewable and non-renewable energy sources

Revised ☐

- A renewable source will continue to be available for our use for a very long time. For example, solar radiation is renewable, but *fossil fuels* are **non-renewable energy sources**.
- **Fossil fuels** (*oil, coal, natural gas*) are formed underground by the action of high pressures and temperatures (in the absence of air) over many millions of years. It is probable that the world's fossil fuel reserves will be significantly *depleted* (reduced) within the next one hundred years.
- Figure 8.1 shows the approximate percentages of various energy sources used currently in the world. The share provided by renewable energy is slowly rising.

> **Key concept**
> A **renewable energy source** is one which is being continuously replaced (over a relatively short timescale) by natural processes.

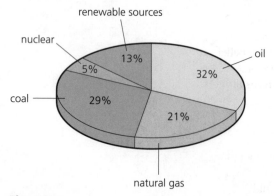

Figure 8.1

Choice of energy sources

- We know from Section 2.3 that the **power** $= \dfrac{\text{energy}}{\text{time}}$ (units: W, kW, MW, GW).
- *Power stations* need to be able to deliver the amount of energy that is needed by the population. This cannot be done without affecting the environment.
- Apart from wanting an energy source to be renewable, so that we do not 'run out of it' (use it all), we may also need to consider the following factors when choosing an energy source:
  - ☐ greenhouse gas emissions (global warming)
  - ☐ health hazards and pollution
  - ☐ ease of transportation of the fuel
  - ☐ energy density or specific energy (explained below)
  - ☐ *efficiency* of energy transfers, usually to electricity (from Section 2.3:
    $$\text{efficiency} = \frac{\text{useful power out}}{\text{total power in}})$$
  - ☐ whether the energy is continuously available 24 hours a day, 7 days a week, 52 weeks a year
  - ☐ (cost).
- In this section we will consider five different energy sources: fossil fuels, nuclear fuels, wind, pumped storage (hydroelectric) and solar power.

> **Expert tip**
>
> There are other important renewable energy sources which are *not* discussed in this course. These include *biomass, geothermal energy* and *wave power*. Biomass is organic material which was recently living, for example wood and oil palm.

■ Specific energy and energy density of fuel sources

- These two concepts are ways of comparing how much energy can be transferred from given amounts of fuels:
- The fact that fossil fuels have high specific energies (typically of the order of $10^7\,\text{J kg}^{-1}$) is a major reason for their widespread use. But the specific energies of nuclear fuels are typically 10^6 times greater.

> **Key concepts**
>
> **Specific energy** = energy transferred from unit mass (unit: J kg^{-1}).
>
> **Energy density** = energy transferred from unit volume (unit: J m^{-3}).

■ Solving specific energy and energy density problems

> **Common mistake**
>
> Energy density (energy per unit volume) and specific energy (energy per unit mass) are often confused and, for solids and liquids, the two concepts convey similar meanings and each have constant values. The specific energy of a gas is constant but its energy density varies with the state of the gas.

QUESTIONS TO CHECK UNDERSTANDING

1.
   a. How much energy is available from 800 kg of coal if its specific energy is $28\,\text{MJ kg}^{-1}$?
   b. How much coal would need to be burned to transfer 1.0 MJ?

2. What is the specific of the chemicals (of mass 10 g) in a torch battery if it can supply a 3.0 W lamp for 2 hours?

3. An oil-fired power station burns 5200 kg of oil every hour. If the specific energy density of the oil is $44\,\text{MJ kg}^{-1}$ and the efficiency of the power station is 35%, what is its output power?

4. The energy density of petrol (gasoline) is $3.4 \times 10^4\,\text{MJ m}^{-3}$.
   a. How much energy is available from 1 cm³?
   b. If the density of petrol $720\,\text{kg m}^{-3}$, what is its specific energy?

Sankey diagrams

- Energy flow **(Sankey) diagrams** can be used to represent the transfer of energy in different processes, or in a series of inter-connected processes.
- *Useful energy* is shown flowing from left to right. The width of each section is proportional to the energy (or power) involved. *Degraded energy* is shown with vertical arrows.
- Every macroscopic process produces degraded energy.

> **Key concept**
>
> **Degraded energy** is energy dissipated into the surroundings as thermal energy which is no longer available to do useful work.

■ Sketching and interpreting Sankey diagrams

■ Figure 8.2 shows a Sankey diagram for the simple one-step process of heating water electrically in a kettle. This process is 90% efficient. If required, the degraded thermal energy could be divided to show the internal energy in the kettle itself.

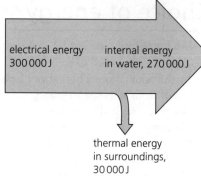

Figure 8.2

QUESTIONS TO CHECK UNDERSTANDING

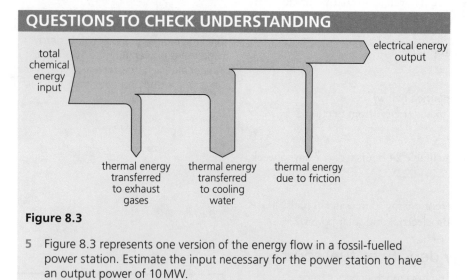

Figure 8.3

5 Figure 8.3 represents one version of the energy flow in a fossil-fuelled power station. Estimate the input necessary for the power station to have an output power of 10 MW.

6 Draw a Sankey diagram to represent the use of a smart phone to play a video game.

Describing the basic features of fossil fuel power stations

Revised ▢

■ Figure 8.4 shows a simplified diagram of a **fossil-fuelled power station**. Note that the steam from the turbine is cooled, condensed and used again.

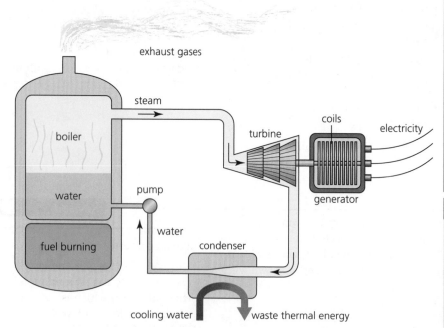

Figure 8.4

Key concept

Thermal energy is used to raise the temperature of water in a boiler and turn it into high-pressure, high-temperature, steam. The steam causes the rotation of **turbines** which are connected to coils of wire in electricity generators. The process of **electromagnetic induction** produces electrical energy as the coils rotate in strong magnetic fields.

Expert tip

Electromagnetic induction is the essential process in the generation of electricity: when a wire is moved across a magnetic field a useful current can be generated. The effect can be increased by winding the wire into coils with many turns and making them rotate very fast in strong magnetic fields. This is explained in more detail in Chapter 11 (for HL students).

- A **turbine** is a device which turns the translational kinetic energy of a fluid into rotational kinetic energy.
- An **electrical generator** is a device which turns mechanical energy (usually of rotation) into electricity.
- Fossil-fuelled power stations have become essential providers of electrical energy in most countries around the world. This is because they rate highly in most of the factors listed above (concerning choice of energy source). But, of course, the world has come to recognize that the burning of fossil fuels contributes to *global warming* (see Section 8.2) and has other harmful environmental effects, such as acid rain.
- The various problems of developing nuclear power and/or renewable sources of energy on the scale necessary to provide for the population of the Earth are considerable. Added to which, the convenience and economy of using the existing infrastructure for fossil fuels, means that reducing the world's dependence on them will take a great many years (despite considerable support for such changes).

Solving problems relevant to energy transformations in the context of fossil fuel power stations

- Calculations may involve power inputs and outputs, related to the efficiency of the power station. The specific energy, or energy density, of a fossil fuel may be needed to relate the amount of fuel used to supply a certain amount of energy (in a given time).

QUESTIONS TO CHECK UNDERSTANDING

7 List at least four reasons why most of the world's electricity is generated from fossil fuels.

8 At what rate must oil be burned in an oil-fired power station to produce an output power of 3.0 GW? (Specific energy = 46 MJ kg^{-1}, efficiency = 39%.)

9 A town of 25 000 people is supplied electricity from one gas-fired power station which has an efficiency of 44%. If the maximum load on the power station occurs when the average power consumption is 1.9 kW per person, what is the greatest rate of gas consumption in the power station (specific energy = 55 MJ kg^{-1})?

10 Make a list of the useful energy transfers which occur in a fossil-fuelled power station.

11 The steam in the boiler in Figure 8.4 is under very high pressure, this raises the boiling point of the water to much higher than 100 °C. Suggest why this is done.

Nuclear power stations

Revised ☐

- We have seen in Section 7.2 that energy can be released by *nuclear fission*. The most widely used example is the fission of uranium-235 that can be induced when a neutron is captured by its nucleus. See Figure 8.5.

- $^{1}_{0}n + ^{235}_{92}U \rightarrow ^{236}_{92}U \rightarrow ^{141}_{56}Ba + ^{92}_{36}Kr + 3^{1}_{0}n$ + gamma ray photons.

- The isotope uranium-235 is the only **fissile** nuclide which is found occurring naturally on Earth in significant quantities. However, this isotope is only about 0.7% of uranium ore (most is uranium-238).

- The products of nuclear fission (called *fission fragments*) have considerable kinetic energy, and if a large number of fissions can be sustained, the nuclear fuel will gain a considerable amount of internal energy and get very hot.

- Figure 8.6 shows one common type of reactor: a *pressurized water reactor* (PWR). Thermal energy is transferred from the fuel to the water which, because it is under high pressure, can get very hot without boiling. Thermal energy (heat) is then exchanged with water in another system using a **heat exchanger**. Steam is created, which drives turbines, as in a conventional power station.

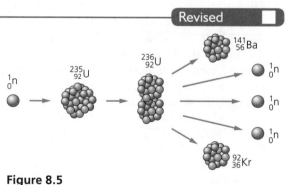

Figure 8.5

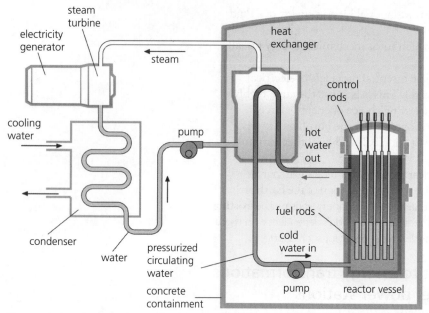

Figure 8.6

Describing the basic features of nuclear power stations

- The pressurized water system in the reactor is exposed to fission products and becomes radioactive. It also acts as the *moderator* of the fission (see below). For safety reasons it is a sealed system within the reactor.
- Each fission reaction, like the one represented by the equation above, produces more neutrons and, under the right conditions, these neutrons can cause further fissions, so that there is the possibility of a **self-sustaining chain reaction** if, on average, one of the neutrons produced in fission goes on to cause further fission.

Sustaining a chain reaction

- The percentage of fissile uranium-235 in a fuel used in a nuclear power station must be increased (above the natural 0.7%) to 3% or more. This process is called **uranium enrichment**. Uranium-238 is a good neutron absorber, so reducing the percentage of it also increases the probability of repeated fissions.
- Neutrons are penetrating particles and many of those produced by fission will pass out from inside the fuel. The ones which remain in and around the fuel must be slowed down to improve their chances of causing further fissions. (Typically from energies of about 1 MeV to about 0.02 eV. This energy is comparable to the energies of molecules in the environment at the prevailing temperature, so that neutrons of this energy are often called *thermal neutrons*.)
- The smaller the mass of the nuclear fuel, the greater the probability of neutrons escaping without causing fission. The minimum mass of nuclear fuel needed to sustain a chain reaction is called its **critical mass**.
- Neutrons are slowed down by collisions with nuclei of light atoms which do not absorb neutrons. This process is known as *moderation*. Graphite and the cooling water (as in Figure 8.6) are widely used as *moderators*.
- The rate of nuclear fission is controlled by *control rods* which are composed of nuclides which are good at absorbing neutrons (boron is commonly used). The rods are moved into or out of the fuel as necessary. See Figure 8.7.

Key concept

The probability of a self-sustaining chain reaction occurring depends on the percentage of uranium-235 in the fuel, the dimensions of the fuel and the speed of the neutrons. **Slow neutrons** are more likely to cause fission of uranium-235.

Expert tip

Since atoms of isotopes have identical chemical properties, the enrichment of uranium has to be a *physical* process which is affected by the very slightly different relative atomic masses of the isotopes (235:238). Gas *centrifuges* are most widely used.

Key concept

The **moderator** in a nuclear power station is a material used to decrease the speed of neutrons.

The rate of reaction is controlled using a neutron absorbing material in **control rods**.

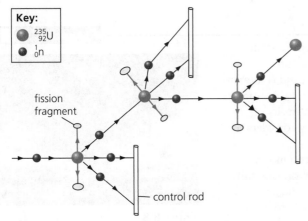

Figure 8.7

Discussing safety issues and risks associated with the production of nuclear power

- Because of its very long half-life (770 million years), radiation from uranium-235 in itself is not considered to be a major concern. However, the radioactive *gas* radon (which has a short half-life) is one of the products in its decay chain, and because it can be inhaled it is strongly linked to lung cancer. This is a major safety concern in the mining and processing of uranium.
- Gamma rays, neutrons and the radiation from fission products are all dangerous and the reactor vessel and its building must prevent the release of these into the environment. People who have to work close to nuclear reactors must be protected and have their radiation exposure monitored regularly.
- Fission products can have very long half-lives. After the fuel becomes of no further use it must be stored for a very long time under conditions which prevent the escape of radiation and radioactive materials from the 'nuclear waste' into the environment.
- Accidents and natural disasters can happen. The design of nuclear power stations must be able to prevent the release of dangerous materials into the environment under extreme circumstances such as a meltdown.
- Radioactive materials could be used as weapons by terrorists. Highly enriched uranium can be used to manufacture atomic bombs.

> **Expert tip**
>
> Many factors contribute to the dangers to health posed by radioactive materials, principally the type of radiation(s) emitted, the physical state of the source and its *activity*. The activity of a pure source depends on the amount of radioactive material and its half-life. A longer half-life means that risks will be present for longer, but the activity will be less (for equal amounts).

Solving problems relevant to energy transformations in nuclear power stations

QUESTIONS TO CHECK UNDERSTANDING

12 a Distinguish between the radioactive decay of uranium-235 and the fission of uranium-235.

   b Suggest why nuclear fission has much greater possibilities as a power source than radioactive decay.

13 Why are waste materials from nuclear power stations dangerous, and how is that risk minimized?

14 Explain why it is much easier to separate isotopes of hydrogen than isotopes of uranium.

15 Explain what would happen if all the pressurized water in the reactor shown in Figure 8.6 were to leak out of the system.

16 Make lists of the principal advantages and disadvantages of nuclear power.

17 The rest mass of a uranium-235 nucleus is 218.9 GeV.

   a Express this mass in unified atomic mass units.

   b If the rest masses of the two nuclides produced in fission are 127.5 GeV and 88.4 GeV, calculate the energy released in each fission, which also produces three neutrons.

   c Calculate the energy that would be released from the fission of 1.0 kg of uranium-235.

   d What would be the average output power of a nuclear power station which transferred this amount of energy from fission over a period of 1 year? Assume its overall efficiency was 30%.

   e Estimate the total amount of enriched uranium needed every year in this power station.

Wind power

- We are using the term **wind generator** to describe an *electricity generator* powered by the wind. Winds are caused by temperature and pressure differences in the air. The original source of the energy in the wind is the Sun.
- There are enormous amounts of energy in the air moving around the surface of the Earth, but that energy is well spread out (due to the low density of air), and the power available in any particular location will vary considerably depending on weather conditions.
- The locations of wind power stations (with many generators), commonly known as *wind farms*, need to be chosen carefully: windy places on bare hills and/or close to the sea are ideal, and offshore sites can be excellent, although they involve extra costs. See Figure 8.8). The rotating blades need to be well above ground/sea level to avoid turbulence and friction. Some people are opposed to wind farms because of noise and visual pollution.
- A single wind generator may be ideal for an isolated home, but the unpredictability of the output means that another energy source will be needed at times of no wind, and/or excess energy can be stored in batteries when there is strong wind.

Figure 8.8

Describing the basic features of wind generators

- Consider Figure 8.9 which shows a 'cylinder' of air about to strike the three blades of a wind generator perpendicularly. This is the most common design for large generators. Wind generators are usually able to rotate so that they always receive the wind perpendicularly.

> **Key concept**
>
> A wind generator transfers the kinetic energy of the air to the rotation of a turbine and then to electricity in an electrical generator.

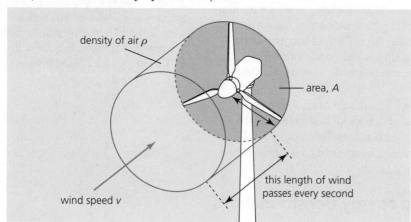

density of air ρ

area, A

r

this length of wind passes every second

wind speed v

Figure 8.9

- Assuming that *all* the kinetic energy of the air is transferred to electricity, the *maximum* theoretical output power can be calculated from $P = \frac{1}{2} A\rho v^3$. (There is no requirement for students to be able to derive this equation.)
- For the sake of simplicity, the use of this equation makes some important assumptions:
  - ☐ the wind is moving perpendicularly to the plane of the blades
  - ☐ the air which strikes the blades will lose all of its kinetic energy
  - ☐ the turbine and generator will be 100% efficient
  - ☐ neighbouring generators will not affect the wind flowing past each other.
- These assumptions may be unrealistic, however the equation does give us a good guide to the power that should be available from a generator of a particular size, especially if we understand that actual efficiencies will vary depending on wind speed and may have a maximum value of around 40% efficiency at transferring available wind energy to electrical energy.

■ Solving problems relevant to energy transformations in the context of wind generators

Expert tip

Clearly, longer blades will transfer more energy (power is proportional to the area swept out and the square of the radius). However, longer blades will be heavier and subjected to greater forces (for the same rate of rotation), so that they (and the rest of the structure) need to be made stronger. Engineers need to decide whether it is more economical to construct more, smaller generators than fewer larger ones.

- The equation shows us that the power is dependent on the wind speed *cubed*. Doubling the wind speed will result in multiplying the power output by a factor of $2^3 = 8$. This example stresses the need for wind generators to receive strong winds. Stronger winds should not imply that the blades need to rotate faster. In fact generators will have an optimum speed of rotation which may seem surprisingly low. Rotational speeds of the blades and the turbine are adjusted to suitable values by gear systems.
- The physics of circular motion (Section 6.1) could also be involved in these calculations.

QUESTIONS TO CHECK UNDERSTANDING

18 a Calculate the output power of a wind generator with blades of length 25 m for a wind speed of 4.0 m s^{-1}. Assume an efficiency of 30%. The density of air is 1.3 kg m^{-3}.

   b Sketch a graph to show how the power output varies up to wind speeds of 16 m s^{-1}.

19 Draw a Sankey diagram to represent the operation of a wind generator.

20 What length of blades is needed to produce an output of 2.0 kW at 25% efficiency for winds of speed 8.0 m s^{-1}?

21 List the advantages and disadvantages of using wind power to generate electricity.

22 Consider a very large wind generator with a radius of 85 m, spinning at 18 revolutions per minute.

   a Calculate:

     i the angular velocity of the blades

     ii the linear speed of a point at the tip of a blade.

   b Consider 10 kg of material as part of the tip of the blade. What is the centripetal force acting on it?

   c Explain why there are practical limits on the diameter and rotational speeds of wind generators.

Hydroelectric power

Revised ☐

- The kinetic energy gained by falling water is used to turn turbines and generate electricity in **hydroelectric power** stations. See Figure 8.10.

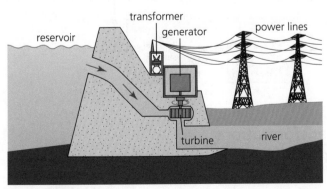

Figure 8.10

- This is a relatively efficient process, typically 90%.
- Water stored in lakes and reservoirs got its gravitational potential energy from the radiant energy of the Sun when it was evaporated to form clouds, which later fell as rain or snow.
- Hydroelectric power stations vary in capacity, but enormous volumes of water are needed to generate sufficient power for large cities. The number of hilly, rainy locations where this is possible is limited. The power stations have the advantage of high efficiencies but there can be many environmental problems.
- The maximum power available can be calculated from potential energy lost by water/time taken, $P = mg\dfrac{\Delta h}{t}$.
- Using this equation, we see that 1 kg of water falling 50 m every second would transfer about 500 W of power. A *specific energy* of 500 J kg^{-1} is very low compared to other energy sources.

> **Common mistake**
>
> The height, Δh, used in these calculations may need to be considered carefully. Under some circumstances the water level may change significantly during the operation of the power station.

Storing electrical energy

- The demand from consumers for electrical power varies considerably, especially between daytime and night-time. An electrical power system must have ways of coping with these variations.
- Current electricity cannot be stored as such, but it can be transferred to another form and then turned back to electricity later. Re-chargeable (secondary) batteries are a common example of this concept, but they are unable to store large amounts of energy. *Pumped storage hydroelectric systems* are the only large-scale, economic way of 'storing electrical energy'.

Describing the basic features of pumped storage hydroelectric systems

- It is better for power systems to keep fuel-powered stations working close to their most efficient outputs, so operators often encourage customers to use power at night by offering lower prices at those times.
- Low-price electricity can be used to pump water up to a reservoir and the profit from selling the hydroelectricity at a higher price at peak times can more than offset the 20% (approximate) loss of energy in the process.
- Hydroelectric power also has the considerable advantage that the output of the systems can be adjusted relatively quickly at times of high demand. This is especially useful if hydroelectricity is used in conjunction with the other less flexible sources, for example wind power or solar power.
- Pumped storage systems need to be in hilly locations. They may be part of a hydroelectric power station, or operate independently, when they add temporary capacity at peak times to a power supply network without using any primary energy source.
- Figure 8.11 shows the Senneca pumped storage system in Pennsylvania, USA.
- Figure 8.12 shows energy transfers in hydroelectric systems.

> **Key concept**
>
> In a **pumped storage hydroelectric system**, at times of low power demand, water is pumped up from a lower reservoir to an upper reservoir and released later to generate hydroelectricity when needed.

Figure 8.11

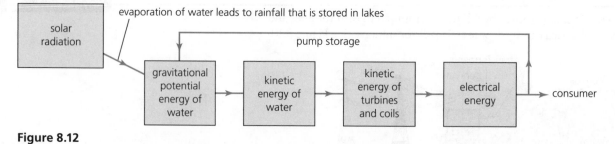

Figure 8.12

Solving problems relevant to energy transformations in the context of pumped storage hydroelectric systems

- Calculations may involve the equation $P = mg\dfrac{\Delta h}{t}$, efficiency and the density of water.

Solar power

Revised ☐

- Radiated energy arrives from the Sun *on the upper atmosphere* at an *intensity* of approximately 1400 W m⁻² (**intensity** = $\frac{\textbf{power}}{\textbf{area}}$). This radiation is mainly visible light and infrared. See Figure 8.13.

- There are wide variations in the intensity *reaching the Earth's surface*. These are due to latitude, weather, seasons and time of day. Devices which use solar energy may be rotatable, so that the incident energy always arrives perpendicularly, so maximizing the total power received by the device.

- Individual solar-powered devices often need to be used with some kind of energy storage system (e.g. batteries or hot water) so that they can continue to be of use at times when less radiation is received.

- The average intensity received over the whole of the Earth's surface (over 24 hours) is about 250 W m⁻².

- There are two principle ways in which we use this renewable energy source directly: *solar heating panels* and *photovoltaic cells* (which are widely called *solar cells*).

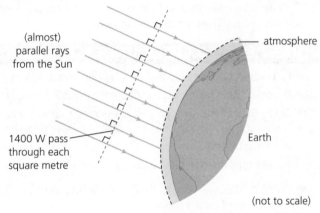

(almost) parallel rays from the Sun

atmosphere

1400 W pass through each square metre

Earth

(not to scale)

Figure 8.13

■ Describing the differences between photovoltaic cells and solar heating panels

- *Solar heating panels* are placed on the roofs of many homes in order to transfer radiant energy from the Sun directly to internal energy in water. The energy is then transferred using a heat exchanger to hot water which can be used in the home. See Figure 8.14. They can be about 50% efficient.

> **Key concept**
>
> In a **solar heating panel** radiant energy from the Sun is transferred to internal energy in water.
>
> In a **photovoltaic cell** the radiant energy produces a potential difference across a semiconductor.

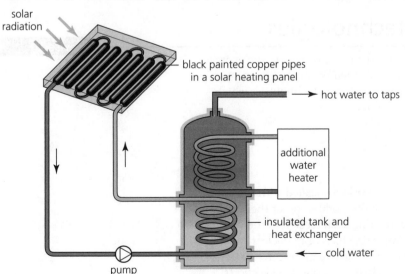

solar radiation

black painted copper pipes in a solar heating panel

hot water to taps

additional water heater

insulated tank and heat exchanger

cold water

pump

Figure 8.14

■ Describing the basic features of solar power cells

- ■ A *photovoltaic (solar) cell* produces a potential difference (voltage) across it when radiant electromagnetic energy releases electrons within the semi-conducting materials of the cell.
- ■ Individual photons within the radiation transfer energy to individual electrons, enabling them to escape from their atoms. This is similar to the *photoelectric effect* which is discussed in Chapter 12 (HL).
- ■ Solar power cells may be used individually (providing a low emf and significant internal resistance), or in small panels, or with very many panels in a large array.
- ■ Cells may be connected in series or parallel to obtain a suitable overall potential difference and internal resistance. Recent developments have seen them used in high power solar-power stations connected to national electricity grids.
- ■ There are many different kinds of solar cells (including thin films). They have a range of different efficiencies, mostly in the range 20–30%.

■ Solving problems relevant to energy transformations in the context of solar power systems

- ■ Questions may involve the intensity of radiation, the area receiving radiation, specific heat capacities and the efficiency of a system. Knowledge of electrical circuits from Chapter 5 may be needed.

QUESTIONS TO CHECK UNDERSTANDING

27 Distinguish between a panel of solar cells and a solar heating panel.

28 Describe how you could investigate the electrical properties of a solar cell.

29 Figure 8.15 shows six solar cells connected together. Each cell has an emf of 0.80 V and an internal resistance of 4.0 Ω. Determine for this arrangement

   a the total emf

   b the total internal resistance.

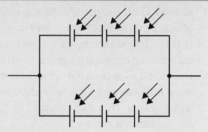

Figure 8.15

30 Figure 8.16 shows solar radiation of intensity 480 W m⁻² incident perpendicularly upon a solar heating panel.

   a Determine the total energy that would be incident upon the panel in one hour.

   b If this energy was used to heat 100 kg of water initially at 15°C and the efficiency was 52%, what temperature could be produced in the water (specific heat capacity = 4200 J kg⁻¹ K⁻¹)?

   c By the time that the sun sets the incident radiation (shown with dotted lines) has fallen in intensity to 220 W m⁻² and the angle of incidence has changed by 40°. Calculate the percentage by which the total energy arriving on the panel every second has changed.

   d Explain why the intensity is less later (or earlier) in the day, than in the middle of the day.

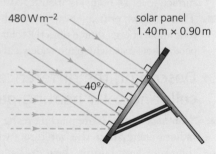

Figure 8.16

New and developing technologies

- ■ Reminder from Section 3.1: Thermal energy is the name we give to the energy transferred from place to place as a consequence of temperature differences.
- ■ Students are expected to be aware of new and developing technologies which may become important during the life of this Guide.

NATURE OF SCIENCE

■ Risks and problem-solving

There have always been very clear and considerable risks associated with the processes of providing the power that the people of the world want for their everyday lives. These include the dangers of extracting fossil fuels from the ground, the possibilities of accidents when transporting fuels or in nuclear power stations, and the consequences of global warming. Recent research and development of renewable energy sources has involved very considerable efforts from scientists around the world to help to solve these problems.

8.2 Thermal energy transfer

Essential idea: For simplified modelling purposes the Earth can be treated as a black-body radiator and the atmosphere treated as a grey body.

Conduction, convection and thermal radiation

■ Reminder: *Thermal energy* is the name we give to the energy *transferred* from place to place as a consequence of temperature differences.

■ If an object contains parts which are at different temperatures, the hottest part contains the particles with the highest average random kinetic energy.

■ When particles collide (interact) energy will be transferred from the particles with greater kinetic energy to particles with less kinetic energy. In this way the thermal energy spreads out until all places have particles with the same average kinetic energy. This familiar process is called **thermal conduction**.

■ Metals are good **conductors** of thermal energy because the energy is mostly transferred by the collisions of *free electrons* between themselves, and between electrons and ions. Most non-metals, liquids and gases are poor conductors of heat (they are good **insulators**).

■ **Convection** currents arise because of the movement of the hotter parts of a gas or liquid when they become less dense (than other parts of the fluid) due to the expansion which occurred when they were heated.

> **Key concept**
>
> The three principle means of thermal energy transfer are *conduction*, *convection* and *radiation*.

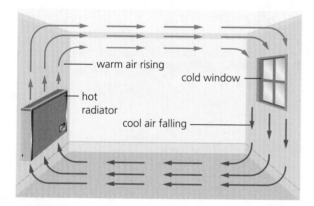

Figure 8.17

■ Figure 8.17 shows a simple example of convection: after thermal energy has been *conducted* from the hot radiator into the air around it, the air expands, becomes less dense and rises. The rising warmer air is replaced by cooler air from below. This encourages the circulation of air as shown in the figure (a *convection current*).

■ The third principal means of energy transfer is by *radiation*. All substances emit electromagnetic radiation due to the motions of charged particles inside them. This is called **thermal radiation** and it is mainly infrared (and visible light if the substance is at very high temperatures). See Section 4.2 for typical wavelengths. Unlike conduction and convection, which need a medium through which to travel, thermal radiation can travel across vacuum. The rest of Section 8.2 is about radiation, the only means by which thermal energy can be transferred to or from the Earth.

> **Expert tip**
>
> Convection is a very common and important phenomenon in liquids and gases. It is the principal means of thermal energy transfer in fluids. For example, an understanding of convection is essential in explaining the weather, ocean currents and movements within the Earth's molten core. Preventing convection currents is often a key feature of good thermal insulation. Water and air are good insulators if convection currents within them can be stopped.

QUESTIONS TO CHECK UNDERSTANDING

31 Outline an experiment which demonstrates that different materials conduct thermal energy at different rates.

32 Explain why the radiator shown in Figure 8.17 might be better placed under the window.

33 Give three examples of good conductors. Explain why good conductors of thermal energy are also good conductors of electricity.

Black-body radiation

- Any object will appear black if very little light is reflected or scattered off its surface. (This is because the light is absorbed).
- Good absorbers of radiation are also usually good emitters of radiation. Conversely, most poor emitters of thermal radiation are also poor absorbers of radiation. White, or shiny, surfaces are a good example.
- A *perfect black body* is an idealized object which emits the greatest possible amount of electromagnetic radiation.

■ Sketching and interpreting graphs showing the variation of intensity with wavelength for bodies emitting thermal radiation at different temperatures

- Figure 8.18 compares spectra of *black-body radiation* at different temperatures.
- A black-body spectrum depends only on temperature, not on the composition or shape of the emitter.

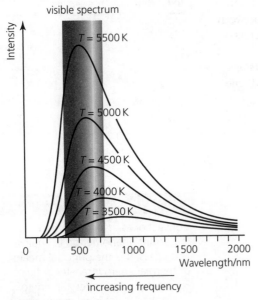

Figure 8.18

> **Key concepts**
>
> A perfect **black body** is an idealized object which absorbs *all* the electromagnetic radiation that falls upon it.
>
> **Black-body radiation** can be defined by a graph which shows, for a given surface temperature, the intensity distribution (*spectrum*) over the different frequencies emitted.

■ Radiated power

- The area under any graph such as that shown in Figure 8.18 is an indication of the total power emitted by the object.
- The total power, P, of the electromagnetic energy emitted from the surface of any perfect black body depends only on its area, A, and temperature, T (not its composition or mass).
- The emitted power can be calculated using the *Stefan–Boltzmann law*.
- The fact that temperature is raised to the power of *four* in this equation indicates that radiated power is very temperature dependent. (For example, doubling the *absolute* temperature of an object increases the emitted power by a factor of $2^4 = 16$.)
- Note that a very hot black body will emit a lot of visible light and will certainly *not* appear black. The Sun is a good example.

■ Emissivity

- A perfect *black body* is an idealized concept. In practice, the total power radiated from all objects will be less than this ideal, although most very hot objects are close to being 'perfect' emitters.
- A perfect black body has an emissivity of one. All objects with surfaces which emit less power (for the same temperature and area) have an emissivity of less than one. They are sometimes described as **grey-bodies**. Most materials have high emissivities.

> **Key concepts**
>
> The *Stefan–Boltzmann law*, $P = \sigma AT^4$, is used to calculate the power emitted by a *perfect black body*.
>
> σ is known as the Stefan–Boltzmann constant $(= 5.67\ 10^{-8}\ \mathrm{W\,m^2\,K^{-4}})$

> **Key concept**
>
> The **emissivity**, e, of a surface is defined as power radiated/power radiated by a black body of the same area and temperature (no units), so that the **Stefan–Boltzmann law** becomes $P = e\sigma AT^4$.

■ Relating surface temperature to the emitted spectrum

- ■ As can be seen in Figure 8.18, the peak of a black-body distribution curve moves to shorter wavelengths (higher frequencies) as the temperature is raised. The value of the wavelength at the peak of the curve, λ_{max}, is inversely proportional to the absolute temperature. This is represented by *Wien's displacement law*.

■ Solving problems involving the Stefan–Boltzmann law and Wien's displacement law

- ■ The Stefan–Boltzmann law can be used to determine the radiated power emitted from any object if its area, temperature and emissivity are known.
- ■ *Wien's displacement law* provides an important and straightforward way of connecting a spectrum to the temperature of the object which emitted it.

> **Key concept**
> **Wien's displacement law:**
>
> $$\lambda_{max}\ (\text{metres}) = \frac{2.90 \times 10^{-3}}{T\ (\text{kelvin})}$$

QUESTIONS TO CHECK UNDERSTANDING

34 Use Figure 8.18 to describe the main differences between the radiation emitted from black bodies at 5500 K and 3500 K.

35 a Determine the power emitted by an object of surface area 120 cm², emissivity 0.74 and surface temperature 100 °C.

   b In what part of the electromagnetic spectrum is this radiation?

36 The filament in a light bulb is 9.6 cm long. If it has a radius of 0.06 mm and is supplied with electrical energy at a rate of 60 W, what will be the surface temperature of the filament? (Assume an emissivity of 1.0.)

37 A surface emits black-body radiation which has a peak frequency of 5.2×10^{14} Hz.

   a In what part of the spectrum is this?

   b What is its temperature?

38 A star of radius 5.0×10^5 km has a surface temperature of 7200 K. Use the Stefan–Boltzmann law and Wien's displacement law to determine:

   a the wavelength at which it radiates maximum power

   b the power emitted by this star.

Radiation from the Sun

Revised ▢

- ■ We can calculate the power emitted from the Sun by using the equation $P = e\sigma A T^4$. (This gives a value of 3.8×10^{26} W.)
- ■ We can assume that this power spreads out equally in all directions with negligible absorption, so that its intensity, I, will decrease according to an inverse square law (Section 4.3). See Figure 8.19.

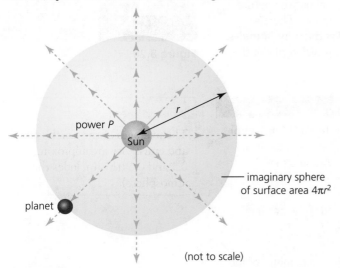

power P

Sun

r

imaginary sphere of surface area $4\pi r^2$

planet

(not to scale)

Figure 8.19

■ The solar constant

- Since intensity, $I = \dfrac{\text{power, } P}{\text{area, } A}$, the intensity of the radiation at different distances, r, from the Sun can be calculated from: $I = \dfrac{P}{4\pi r^2}$.

- This equation can be used to determine the intensity of radiation falling on an area above the Earth's atmosphere which is perpendicular to the direction in which the radiation is travelling. Its value is $1.36 \times 10^3 \, \text{W m}^{-2}$ and it is called the *solar constant*, although there are slight variations mainly due to cyclical changes within the Sun itself, as well as small changes to the Sun–Earth distance. See Figure 8.13.

> **Key concept**
>
> The **solar constant** is the value of the intensity of the radiation from the Sun reaching the top of the Earth's atmosphere.

> **Expert tip**
>
> The Stefan–Boltzmann law and Wien's displacement law are widely used in astronomy (Option D). The power emitted by a star (called its luminosity), the received intensity and the star–Earth distance are easily connected by the inverse square law. The temperature of a star can be determined from its emission spectrum.

■ Albedo and emissivity of the Earth

- The (equilibrium) temperature of the Earth's surface and surrounding atmosphere is discussed later in this section. In broad terms, the temperature depends on:
  - ☐ the radiated *power received* from the Sun (as above)
  - ☐ the percentage which is reflected straight back into space (represented by the Earth's **albedo**: see right)
  - ☐ the power radiated away from the Earth's surface and its atmosphere (dependent upon the their *emissivities*).
- The *emissivity* of the Earth's *surface* is generally greater than 0.9 and we may often consider that it is effectively a black body, but the effect of clouds and the atmosphere reduces the average value of the Earth *and* its atmosphere to an emissivity usually assumed to be about 0.6. The atmosphere may be said to act as a grey body.
- Surfaces which scatter and reflect radiation well (and absorb poorly), like snow and ice, are said to have a high *albedo*.
- The average albedo of the Earth and its atmosphere is usually assumed to be about 0.3. The high percentages of water, snow and ice on the Earth's surface, together with cloud coverage, have the greatest effects on this overall average. Figure 8.20 represents the *approximate* values for the albedos of water, ice, snow and land/vegetation.
- All albedo values for various materials should be considered as just guides because they can change with a more detailed knowledge about the nature of the surface, and with the angle of incidence of the radiation. There are variations of albedo during the day, during the year and at different latitudes. Albedo values are also dependent to some extent on the wavelengths of the radiation.

> **Key concept**
>
> Albedo (α) = total scattered power/total incident power (there are no units because it is a ratio).

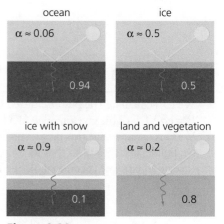

Figure 8.20

> **Key concept**
>
> The averages of the Earth's albedo and emissivity are approximately 0.3 and 0.6 (this includes the atmosphere).

QUESTIONS TO CHECK UNDERSTANDING

39 Use the value of the solar constant and the Earth–Sun distance (1.5×10^{11} m) to confirm the radiated power emitted from the Sun.

40 Calculate the total power radiated away from the Earth. Assume the average surface temperature is 16 °C, the radius is 6.4×10^6 m and its average emissivity is 0.61.

41 A lake has a surface area of 3.4 km². In the middle of the day the average albedo is 0.08, by the late afternoon it has risen to 0.13.

   a Suggest why the value of the albedo changed.

   b If the incident power from the Sun is 235 W on each square metre of the lake's surface at noon, at what rate is energy absorbed by the lake?

   c By the late afternoon the incident power is halved. What is the intensity of the reflected radiation at that time?

The greenhouse effect

- The Earth's atmosphere keeps the planet warmer than it would be without an atmosphere. This is called the (natural) *greenhouse effect*. It is important to distinguish between the *natural greenhouse effect* (which is beneficial) and the *enhanced greenhouse effect* (which is contributing to global warming: see later).
- Consider the Earth (radius r) as it is now, but without an atmosphere: total power received $= \pi r^2 \times 1360\,\text{W m}^{-2}$. Power radiated away, $P = e\sigma(4\pi r^2)T^4$. Assuming (for the sake of simplicity) an emissivity of 1 and an albedo of 0.3, this equation indicates that the equilibrium surface temperature of the Earth *without an atmosphere* would be approximately 250 K. The actual *average* surface temperature of the Earth *with* an atmosphere is believed to be 287 K. The increase is due to the greenhouse effect.

> **Key concept**
>
> **Greenhouse effect**: Infrared radiation emitted by the Earth is partially absorbed by certain gases in the atmosphere. The radiation is then re-emitted in random directions. Some of it is then re-absorbed by the Earth's surface, keeping it warmer than it would be without an atmosphere.

■ Describing the effects of the Earth's atmosphere on mean surface temperatures

- Now consider the effect of an atmosphere: about 30% of the radiation received from the Sun is reflected and about 20% is absorbed by the atmosphere. The remaining 50% (approximately) is absorbed by the Earth's surface. However, when infrared radiation is emitted from the Earth's surface, a much higher percentage is absorbed in the atmosphere. This is because the emitted radiation has longer wavelengths than the incoming radiation from the Sun. The energy is then re-emitted by the atmosphere, but in random directions, so that some of the radiation which was travelling away from the Earth is radiated back towards the Earth. This results in a higher equilibrium temperature of the Earth's surface (compared to an Earth without an atmosphere).

- Figure 8.21 indicates what happens to the total solar radiation incident on the Earth and its atmosphere. (Note that the height of the atmosphere has been much exaggerated for the sake of clarity.)

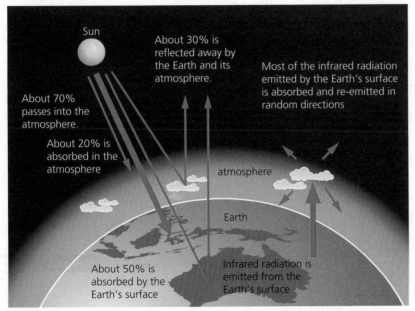

Figure 8.21

- The reason for the absorption in the atmosphere of some radiation from the Earth's surface, is that the atoms within all gas molecules oscillate (see Figure 8.22 for an example which shows possible oscillations in a carbon dioxide molecule) and, if the frequency of these oscillations is the same as the frequency of the radiation, then energy is transferred from the radiation to the molecules (increasing their kinetic energies of oscillation).

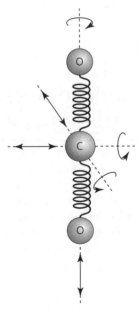

Figure 8.22

■ The radiation which can be absorbed by this effect is only found within the *infrared* section of the electromagnetic spectrum. These wavelengths are common in the radiation *emitted from the Earth*, but less common in the radiation emitted by the *much hotter* Sun (consider again Figure 8.18). The effect of the atmosphere is similar to that of the glass in a *greenhouse*: the shorter wavelengths from the Sun can pass through, but some of the longer wavelengths emitted from the plants, etc. inside cannot pass back out through the glass.

■ The gases which contain molecules capable of absorbing infrared radiation are commonly known as the **greenhouse gases**. The most common ones are *water vapour, carbon dioxide, methane* and *nitrogen dioxide*. All these gases occur naturally, but the last three are now present in the atmosphere in increased amounts because of human activities.

> **Key concept**
>
> Greenhouse gases in the atmosphere (for example carbon dioxide) absorb infrared radiation because their molecules oscillate at the same frequencies as the radiation.

> **Expert tip**
>
> When any system which can oscillate is disturbed by another external oscillation, energy can be transferred to that system if its natural frequency of oscillation is the same as the external frequency. This effect is called *resonance* and it is dealt with in Option B (HL). Molecules of greenhouse gases resonate in the infrared section of the electromagnetic spectrum.

■ Figure 8.23 shows a more detailed representation of the energy flows to and from the Earth and its atmosphere. (The numbers are approximate and need not be remembered.)

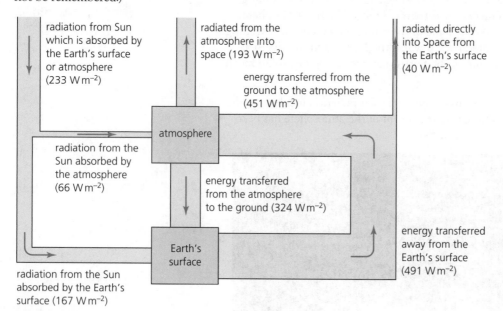

radiation from Sun which is absorbed by the Earth's surface or atmosphere (233 W m^{-2})

radiated from the atmosphere into space (193 W m^{-2})

radiated directly into Space from the Earth's surface (40 W m^{-2})

energy transferred from the ground to the atmosphere (451 W m^{-2})

atmosphere

radiation from the Sun absorbed by the atmosphere (66 W m^{-2})

energy transferred from the atmosphere to the ground (324 W m^{-2})

Earth's surface

energy transferred away from the Earth's surface (491 W m^{-2})

radiation from the Sun absorbed by the Earth's surface (167 W m^{-2})

Figure 8.23

The enhanced greenhouse effect and global warming

Revised ☐

■ Human activities, especially burning fossil fuels, have increased the concentration of greenhouse gases in the atmosphere (although the percentage of water vapour has not changed significantly). This has almost certainly led to an *enhanced greenhouse effect* and **global warming**.

■ As evidence for the enhanced greenhouse effect, there is a reasonably close correlation between mean global temperatures and the amount of carbon dioxide in the atmosphere, (i) over hundreds of thousands of years (using ice core research) and (ii) in recent years, since more accurate measurements have been possible.

> **Key concept**
>
> **Enhanced greenhouse effect:** The Earth's atmosphere has been slightly changed because of human activities. Greater concentrations of greenhouse gases have resulted in slightly less infrared radiation being radiated into space. As a consequence, it is believed that global temperatures are slowly rising.

- As a result of human activities, *more* infrared energy is absorbed by the *increased* concentration of greenhouse gases, so that *more* energy is re-radiated back to the Earth's surface, resulting in a higher equilibrium temperature.
- Scientists believe that even a small increase in average temperature will lead to climate changes and subsequent slight changes to the Earth's average albedo and emissivity, which are then likely to lead to further temperature increase (*positive feedback*).
- To reduce global warming, we should try to reduce our use of energy. Much greater use of renewable energy resources and nuclear power will also decrease the amount of fossil fuels that will need to be burned. The efficiency of power stations can be slightly improved (using natural gas is the most efficient) and combined heat and power stations can use the thermal energy that is usually transferred wastefully to the surroundings for heating nearby buildings.

Common mistake

It is important to distinguish between the *natural* greenhouse effect and the *enhanced* greenhouse effect which has man-made origins.

■ Energy balance in the Earth surface–atmosphere system

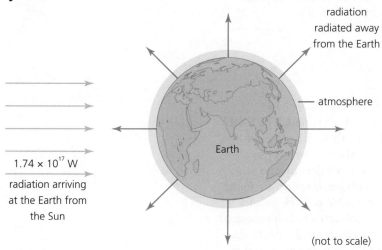

Figure 8.24

- A planet, like the Earth, will stay in *energy balance* (thermal equilibrium) if the average power of the radiation that it receives, equals the average power it radiates into space.
- Using the equation: **solar constant × cross-sectional area of Earth × (1 − albedo, α) = $e\sigma AT^4$** confirms an average surface temperature of the Earth of about 290 K (using $e = 0.60$ and $\alpha = 0.30$), which is close to the accepted value of 287 K.
- This equation is a simple, but informative, model of a very complex situation. It can be used to make basic predictions of the consequences if global warming leads to significant changes in the Earth's average emissivity and/or albedo.
- Computer programs can be used to develop much more detailed climate models and predict possible future outcomes. However, the situation is extremely complicated, with many interconnected variables. Despite considerable international efforts and expertise, there is still no widespread consensus about the detailed consequences of global warming.

> ### Key concept
> The Earth's temperature adjusts so that that the power of the infrared energy radiated away into space equals the power of the incident solar radiation. This under-lying principle is an important basis for models of the future climate.

Solving problems involving albedo, emissivity, solar constant and the Earth's average temperature

QUESTIONS TO CHECK UNDERSTANDING

42 Make a calculation which confirms that the average temperature of the Earth without an atmosphere would be about 250 K.

43 Explain how the natural greenhouse effect increases the surface temperature of any planet which has an atmosphere.

44 a Explain what is meant by saying that water vapour is a 'greenhouse gas'.

   b Explain why scientists are *not* worried that water will contribute to the *enhanced* greenhouse effect.

   c Name the three principal greenhouse gases (other than water vapour).

   d Choose *one* of these three gases and explain why there is an increased amount of it in the atmosphere.

45 Explain how the *enhanced* greenhouse effect can lead to global warming.

46 a Use the equation in the previous section to confirm that it predicts an average surface temperature for the Earth of 290 K (using $e = 0.60$ and $\alpha = 0.30$).

   b Determine the temperature rise that this simple model predicts if, for example, the Earth's emissivity and albedo both fell by 5%.

NATURE OF SCIENCE

■ Simple and complex modelling

The kinetic theory of gases (Section 3.2) is a model that can be applied successfully to relatively small amounts of gases in closed containers. The Earth's atmosphere is also a gaseous system, but one which is vastly larger and more complex. Meteorologists use computer modelling to predict the weather in a particular location with reasonable accuracy up to about ten days in advance. However, predicting the climate of an entire planet with any certainty for many years in the future is a very difficult, if not almost impossible task. However, climate modelling is a problem which has understandably attracted an enormous amount of scientific attention in recent years and, with the availability of better data and faster processing, together with international collaboration, long-term climate models are believed to have become more consistent and reliable. We will have to wait to see how accurate they are!

9 Wave phenomena

9.1 Simple harmonic motion

Essential idea: The solution of the harmonic oscillator can be framed around the variation of kinetic and potential energy in the system.

The defining equation of SHM

- SHM has already been introduced in Section 4.1.
- In this section we will take this further by putting a constant into the equation, enabling a mathematical approach which will allow the calculation of displacements, velocities and energies at any chosen time:
- Angular velocity, ω, (first met in Section 6.1) is a key feature in describing all simple harmonic motions. This may seem strange at first because, for example, a mass oscillating with SHM on the end of a spring is clearly *not* rotating in a circle. However, circular motion and oscillations are mathematically very similar. Figure 9.1 shows why: as a particle moves in a circle with constant speed, its projection onto the diameter, P, oscillates back and forth with simple harmonic motion.
- Reminder from Section 6.1: angular velocity, $\omega = \dfrac{2\pi}{T}$ or $\omega = 2\pi f$. These equations enable us to quickly convert observed time periods, or frequencies, into the equivalent angular velocities in $\text{rad}\,\text{s}^{-1}$ (as needed in the equations later in this section).

> **Key concepts**
>
> **Simple harmonic motion** occurs when a (restoring) force produces an oscillation in which the acceleration is proportional to the displacement, *but in the opposite direction*: $a \propto -x$.
>
> Putting in a constant (ω^2):
> $a = -\omega^2 x$.

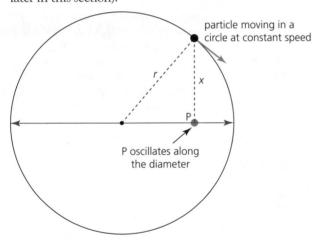

particle moving in a circle at constant speed

P oscillates along the diameter

Figure 9.1

- A graph of displacement–time can be used to fully represent any SHM, as shown in Figure 9.2.

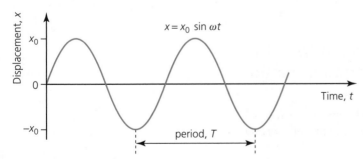

$x = x_0 \sin \omega t$

period, T

Figure 9.2

Equations for displacement and velocity

- The graph in Figure 9.2, which starts at $t = 0$ with zero displacement, is sinusoidal in shape and can be represented by the equation: $x = x_0 \sin \theta$ or $x = x_0 \sin \omega t$ (remembering from Section 6.1 that $\omega = \frac{\theta}{t}$)
- The velocity of the oscillating object at any time can be determined from the gradient of the displacement–time graph (Section 2.1). If an acceleration is required, it can be determined from the gradient of a velocity–time graph.
- The velocity can be determined directly using the equation $v = \omega x_0 \cos \omega t$ (for an oscillation which has zero displacement at $t = 0$).
- Figure 9.3 shows displacement, velocity and acceleration graphs on the same axes for easy comparison. A velocity–time graph is $\frac{\pi}{2}$ out of phase with a displacement graph of the same oscillation. The acceleration graph is $\frac{\pi}{2}$ out of phase with the velocity graph.

> **Key concept**
>
> Graphs showing the variations of displacement, velocity and acceleration with time during SHM are all sine or cosine waves and can be represented by trigonometric equations. The graphs are out of phase with each other.

> **Common mistake**
>
> Be sure to check the displacement at time $t = 0$ before deciding which equation to use for subsequent displacement or velocity.

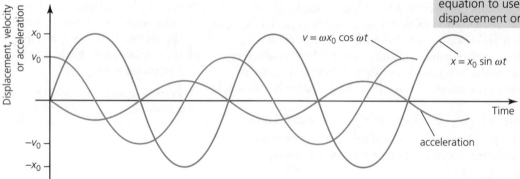

Figure 9.3

- Figure 9.4 represents the graphs for a SHM which *began at time $t = 0$ with maximum displacement*. Note the corresponding equations for displacement and velocity: $x = x_0 \cos \omega t$ and $v = -\omega x_0 \sin \omega t$

> **Expert tip**
>
> The velocity equations can be obtained from the displacement equations by the mathematical process of *differentiation*. But this is not required by this course.

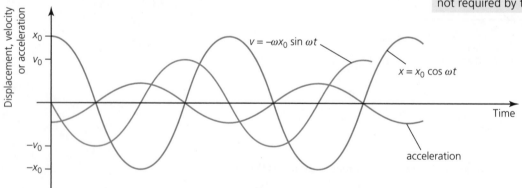

Figure 9.4

▪ Solving problems involving acceleration, velocity and displacement during simple harmonic motion, both graphically and algebraically

- If the angular velocity and amplitude of a SHM are known, the equations highlighted above can be used to calculate the displacement, velocity and acceleration at any time.
- More commonly, we may want to know what the velocity is for a given displacement, then we can use the equation: $v = \pm \omega \sqrt{\left(x_0^2 - x^2\right)}$

QUESTIONS TO CHECK UNDERSTANDING

1 A student determines that a pendulum completes 20 oscillations in 15.81 s.

   a What is: **i** the period, **ii** the frequency of this oscillator?

   b Calculate the angular velocity of the oscillations of the pendulum.

   c If the pendulum behaves as a simple harmonic oscillator with amplitude 8.3 cm, determine its displacement 2.5 s after it was released from its maximum displacement.

2 A simple harmonic oscillator has an angular velocity of 74 rad s^{-1}.

   a What is its frequency?

   b What is the magnitude of the acceleration when it has a displacement of 1.0 cm?

   c Calculate the maximum velocity of the oscillator if it moves with an amplitude of 3.3 cm.

   d Explain why there are two possible answers to part **c**.

3 Figure 9.5 shows the displacement–time graph for a simple harmonic oscillator.

   a What is its angular velocity?

   b Write down an equation to represent this graph.

   c Determine the velocity of the oscillator after 2.5 s.

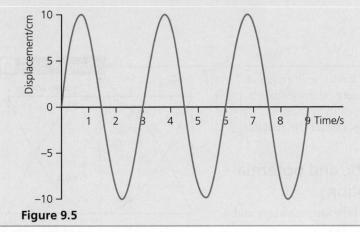

Figure 9.5

The simple pendulum and a mass–spring system

Revised ☐

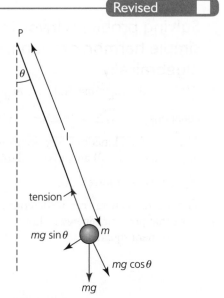

Figure 9.6

- These two oscillators are the most commonly investigated in school laboratories because they approximate well to SHM and also because they have time periods which are easily observed and measured.

- We are surrounded by objects which oscillate and the analysis of more complicated oscillators often begins by making comparisons with one of these two basic models. For example, oscillations within molecules (for example, in greenhouse gases) are often compared to masses vibrating on springs.

- A **simple pendulum** can be considered as a point mass oscillating on the end of a string which has negligible mass and does not stretch. Metal spheres on the end of thin strong strings approximate well to simple pendulums. Figure 9.6 shows a diagram of a simple pendulum. There are two forces acting on the mass, m (the pendulum 'bob'): its weight, mg, downwards and the tension in the string.

- The mass is not in equilibrium and it will accelerate back towards the vertical. The weight, mg, is conveniently split into two components: $mg\sin\theta$, which provides the restoring force, and $mg\cos\theta$ along the line of the string, which is equal and opposite to the tension.

- We know from Newton's second law that $F = ma$, which can be applied to the restoring force, so that: $ma = -mg\sin\theta$. A minus sign has been included here to represent the fact that the force acts in the opposite direction to increasing displacement.

- For small angles (up to about 10°), $\sin\theta \approx \theta = \frac{x}{l}$, where x is the displacement, so that $a = -\left(\frac{g}{l}\right)x$.
- Comparing this equation for acceleration to the defining equation for SHM ($a = -\omega^2 x$), leads to $\omega^2 = \frac{g}{l}$.

- Since $\omega = \frac{2\pi}{T}$, the equation for the period of a simple pendulum is $T = 2\pi\sqrt{\frac{l}{g}}$.
- The period does not depend on the amplitude so that the oscillation is *isochronous*. The period is also independent of the mass: all *simple* pendulums of the same length have the same period, regardless of their mass or amplitude.
- Consider again Figure 4.1, which shows a simple **mass–spring system**. From Section 2.3 we know that if the force on any spring system is proportional to the displacement, we can describe it using the concept of *force constant*, $k = \frac{F}{\Delta x}$.
- Restoring force from the springs, $F = -kx = ma$, so that $a = -\frac{kx}{m}$. A minus sign has been included here to represent the fact that the force acts in the opposite direction to increasing displacement.
- Comparing this equation for acceleration to the defining equation for SHM ($a = -\omega^2 x$), leads to $\omega^2 = \frac{k}{m}$.
- Since $\omega = \frac{2\pi}{T}$, the equation for the period of a simple mass–spring system is $T = 2\pi\sqrt{\frac{m}{k}}$.

Key concept

By considering the restoring forces acting on them, equations can be developed for the periods of two basic and important oscillators: a simple pendulum and a mass–spring system.

Energy changes

- All mechanical oscillators continuously interchange kinetic energy, E_K, and some form of potential energy, E_p (and some of the energy is dissipated into the surroundings as thermal energy). This was discussed in Section 4.1. We will now discuss this in more detail by including equations for the energies involved.

Describing the interchange of kinetic and potential energy during simple harmonic motion

- Figure 9.7 from Section 4.1 is repeated here. It shows how kinetic energy and potential energy of perfect SHM vary with *displacement*.
- Note that for perfect SHM there is no energy dissipation, so that the total energy, E_T, remains constant: $E_T = E_K + E_p$
- Figure 9.8 shows how these energies change with *time*.

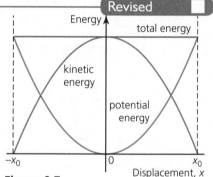

Figure 9.7

Solving problems involving energy transfer during simple harmonic motion, both graphically and algebraically

- We know that $E_K = \frac{1}{2}mv^2$ and $v = \pm\omega\sqrt{\left(x_0^2 - x^2\right)}$, combining these two equations gives us $E_K = \frac{1}{2}m\omega^2(x_0^2 - x^2)$.
- The maximum kinetic energy occurs when potential energy is zero (when displacement, $x = 0$) and it is equal to the total energy of the oscillator:

$E_{K\,max} = E_T = \frac{1}{2}m\omega^2 x_0^2$.

- The potential energy can be determined from $E_p = E_T - E_K$: $E_p = \frac{1}{2}m\omega^2 x^2$.
- Note that potential energy is shown by this equation to be proportional to the displacement squared.

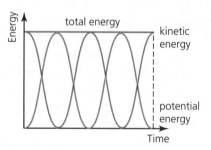

Figure 9.8

QUESTIONS TO CHECK UNDERSTANDING

4 a Explain why the periods of a simple pendulum and a mass–spring system do not depend on the amplitude of their oscillations.

   b What length of a simple pendulum will produce a period of exactly 1.0 s?

   c Sketch a graph to show how the period of a simple pendulum depends on its length.

5 A force of 5.6 N stretches a spring by 4.9 cm.

   a What is the force constant of this spring?

   b A mass of 900 g was made to oscillate vertically on the end of the spring. What is the time period of this oscillation?

   c A stopwatch was started when the mass was at its lowest point, which was 2.9 cm below its equilibrium position. Determine the displacement and velocity of the mass 0.80 s later (assume the motion is simple harmonic).

6 Figure 9.8 represents the energies of a perfect simple harmonic oscillator. In reality many practical oscillators will dissipate significant amounts of energy.

   a Suggest a possible reason for this dissipation.

   b Sketch how the potential energy graph would appear if about 50% of the total energy was dissipated into the surroundings every oscillation.

7 A 2.8 g mass is oscillating with an angular velocity of 73 rad s^{-1}. This is SHM and the amplitude is 2.0 cm.

   a Calculate the total energy of the oscillator.

   b Determine the potential energy when the displacement is 0.38 cm.

   c Calculate the kinetic energy when the displacement is 1.3 cm.

   d What is the maximum speed of the mass?

NATURE OF SCIENCE

■ Insights

The theory of the basic simple harmonic oscillator (as discussed in this section) is the starting point for the analysis of a large number of more complex oscillating systems (molecular and electronic, as well as mechanical). The simplicity provided by basic models which can be easily visualized (pendulums and masses on springs) can provide the necessary insight for more advanced modelling.

9.2 Single-slit diffraction

Revised ☐

Essential idea: Single-slit diffraction occurs when a wave is incident upon a slit of approximately the same size as the wavelength.

■ The wave property of **diffraction** was introduced in Section 4.4, where it was noted that diffraction effects are greatest when the gap (*aperture*), or obstacle, has a size which was similar to the wavelength of the waves.

■ Light waves have a small wavelength (typically 5×10^{-7} m), so if we wish to observe the diffraction of light, a very small aperture must be used.

■ The diffraction pattern produced by light passing through a single narrow slit was briefly discussed in Section 4.4. In this section that pattern will be described in more detail and explained using the concept of interference.

> **Key concept**
> A single slit diffraction pattern for monochromatic light consists of a series of light and dark bands (fringes). The central band is twice the width of the others and very much brighter.

The nature of single-slit diffraction

Revised ☐

■ Figure 9.9 shows a typical experimental arrangement used for observing diffraction caused by a narrow slit. In this example, *monochromatic* laser light is being used, although it is not essential. A *parallel-sided rectangular slit* produces the simplest diffraction pattern (the pattern from a circular hole would be a series of rings).

> **Key concept**
> Single slit diffraction patterns can be explained in terms of the interference of light coming from different points across the width of the slit.

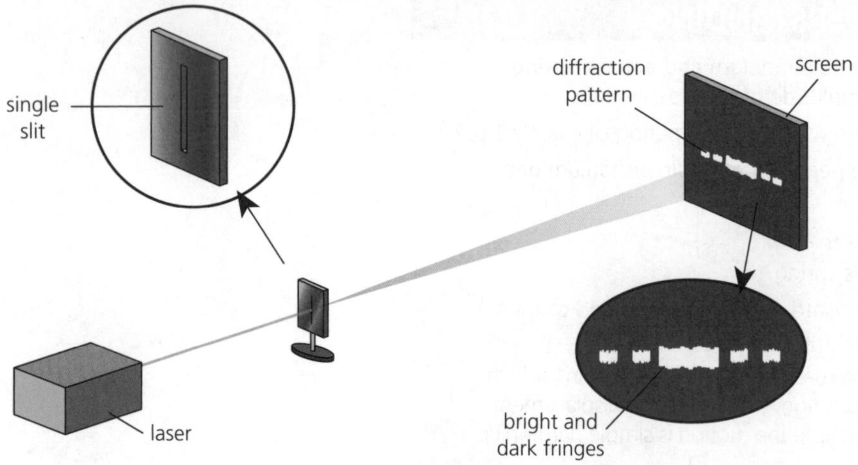

Figure 9.9

■ Figure 9.10 compares the patterns that can be seen when using monochromatic light (Figure 9.10a) and white light (Figure 9.10b, discussed later). The regions of light and dark are sometimes called *bands* or *fringes*.

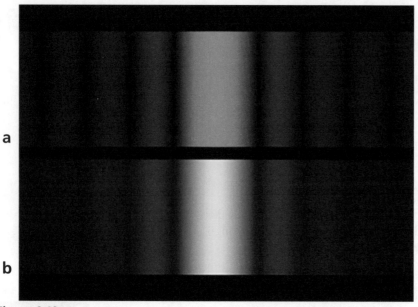

Figure 9.10

■ If the width of the slit was reduced the pattern would spread out and get dimmer.

Qualitatively describing single-slit diffraction patterns produced from a range of monochromatic light frequencies

■ Figure 9.11 shows how the variation in intensity of the light received on a screen can be represented graphically. (The values given for the angles are explained below.)

■ This graph can be confused with a similar one for the pattern seen with two (or more) slits discussed in Section 9.3, so it is important to note the distinguishing features of the *single-slit diffraction pattern*: the central maximum is twice as wide as all the others, and the intensity of the other peaks decreases significantly with distance from the centre of the pattern: if the central maximum has an intensity of I_0, the first has an intensity of about $\frac{I_0}{20}$ and the second aproximately $\frac{I_0}{50}$. (Figure 9.11 is not drawn to scale.)

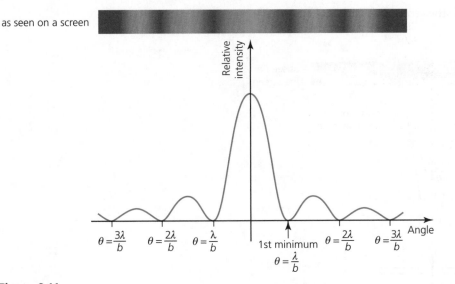

as seen on a screen

Figure 9.11

Explaining the single-slit diffraction pattern

- If light was incident on a gap with a width equal to the wavelength, diffraction would cause the waves to spread out equally in all directions (as described for waves generally in Section 4.4). However with light, slits much wider than a wavelength must be used in order to allow sufficient light to pass through them. A narrow slit of width only 0.1 mm is still about 200× wider than one wavelength of light.
- In the *single-slit diffraction pattern*, light waves clearly do *not* spread equally in all directions. In order to explain this, waves diffracting away from the slit can be considered as a series of *wavelets* originating from imagined separate sources in the slit which are about one wavelength apart from each other. (So that, a slit 200× wider than one wavelength would be considered to have 200 sources of wavelets.)
- These wavelets *superpose* with each other to produce an *interference* pattern, although it is still called the *diffraction* pattern of a single slit. This is explained as follows:
- Figure 9.12 shows two wavelets from the edges of a slit of width b. If these two wavelets interfere constructively at an angle θ, then the *path difference* between them, $b\sin\theta$ (as shown) will equal one wavelength, λ (as explained in Section 4.4).

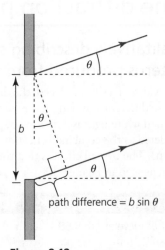

path difference = $b \sin \theta$

Figure 9.12

- However, if the two extreme wavelets interfere constructively then waves from halfway across the slit must interfere destructively with them because the path difference = $\frac{\lambda}{2}$. This means that wavelets from each point in the lower half of the slit can be paired off with wavelets from the top half of the slit. In this way the situation in Figure 9.10 becomes the condition for overall *destructive* interference.

Determining the position of first interference minimum

- As explained above $\lambda = b\sin\theta$ is the condition under which the first minimum of the pattern is formed. For the small angles that will be involved in this experiment, $\sin\theta \approx \theta$, so that the equation $\theta = \frac{\lambda}{b}$ predicts the angle (in radians) of the first interference *minimum* of the pattern produced by light of wavelength λ passing through a narrow slit of width b.
- Other minima occur at angles which are multiples of θ. Note the angles labelled on Figure 9.11.
- In the IB Physics course, calculations will only involve determination of the position of the *first minimum* of the pattern.

> **Key concept**
>
> The first diffraction minimum of a single slit diffraction pattern occurs at an angle $\theta = \frac{\lambda}{b}$

> **Expert tip**
>
> Although this section concentrates on visible light, whenever waves of any kind are emitted or received, we may consider that they are passing through an aperture and will therefore be diffracted.

■ Figure 9.13 relates the angle θ to the geometry of a typical experiment.

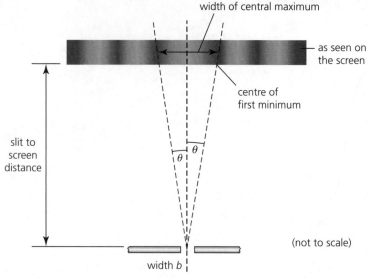

Figure 9.13

Describing the effect of slit width on the diffraction pattern

Revised

■ Qualitatively describing single-slit diffraction patterns produced from white light

■ If the width of a slit is reduced, the diffraction pattern gets wider and dimmer.
■ Because different wavelengths are sent in different directions by the same slit, when white light is diffracted by a narrow slit the pattern seen can be considered as the combined effect of many overlapping monochromatic patterns. The overall effect is that the interference bands are edged with colour, as seen in Figure 9.10b.

Expert tip

When white light passes through a narrow slit, some separation of colours will be seen within the bands (fringes).

QUESTIONS TO CHECK UNDERSTANDING

8 Light of wavelength 5.68×10^{-7} m (568 nm) is passed through a slit of width 8.4×10^{-2} mm.

   a Calculate the angle at which the first minimum is formed: **i** in radians, **ii** in degrees.

   b If the pattern is seen on a screen which is 3.42 m from the slit, what is the width of the central maximum?

9 a Sketch a labelled graph to show how the intensity of light of wavelength 476 nm varies with angle (rad) after passing through a narrow slit of width 5.9×10^{-5} m. Include the central maxima and two others on either side.

   b Add a second graph to your axes to show the effect of halving the slit width.

10 Make a copy of the graph shown in Figure 9.11 (which may be assumed to represent red light). On the same axes draw a graph to represent the diffraction of blue light passing through the same slit.

11 A teacher wished to demonstrate the diffraction of light from a helium laser through a narrow slit of width 0.052 mm. What is the minimum distance between the slit and the screen in order for the central maximum to have a width of 4.0 cm? (Wavelength = 6.12×10^{-7} m.)

■ Development of theories

The diffraction of light was first noticed hundreds of years ago, but at that time no theory of light was able to explain the observations. The wave theory of light provided a good explanation of diffraction, but the more recent quantum theory of radiation necessitates that the wave theory is amended (not required in this chapter).

9.3 Interference

Revised ☐

Essential idea: Interference patterns from multiple slits and thin films produce accurately repeatable patterns.

■ The interference of *coherent* waves has already been discussed in Section 4.4. The conditions for *constructive and destructive interference* were explained and practical examples given for sound, microwaves and light.
■ We will now look at the *double-slit interference of light* (**Young's experiment**) in more detail and then apply that knowledge to the use of *multiple slits*.

Young's double-slit experiment

Revised ☐

■ This experiment is of historical importance because it provided the first evidence that light travelled as waves (only waves can interfere).
■ Two coherent sources are provided by using two narrow slits to split the wavefronts from a single source of light.

■ Investigating Young's double-slit experimentally

■ The experiment is most easily performed with laser light in a darkened room, see Figure 9.14. The two narrow slits are close together and placed in front of the laser and about 2 m or more from a screen on which the interference pattern is observed.

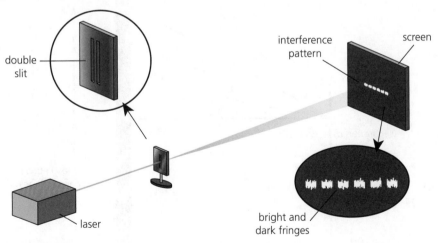

Figure 9.14

■ Measurements are made of the distance between the centres of the two slits, d, the perpendicular distance between the slits and the screen, D, and the distance between the centres of neighbouring fringes seen on the screen, s. (Usually the separation of a known number of fringes will be measured.)
■ The wavelength, λ, of the light can be determined from the equation given previously in Section 4.4: $s = \dfrac{\lambda D}{d}$. The origin of this equation is explained below.

■ Qualitatively describing two-slit interference patterns, including modulation by one-slit diffraction effect

- ■ Figure 9.15 compares the interference patterns of monochromatic light and white light passing through the same two slits.
- ■ These patterns may appear similar to single-slit diffraction patterns (Figure 9.10), but the differences should be noted: the fringes in the centre of the (monochromatic) interference pattern are all approximately the same width and intensity. Note that m*odulation* of two-slit interference patterns (by the one-slit diffraction effect) is not apparent in this figure. Modulation is explained later.

Figure 9.15

■ Explaining the double-slit equation

- ■ Figure 9.16 shows two parallel light rays representing waves emerging from double narrow slits which have a distance *d* between their centres. If waves travelling in this direction interfere constructively, we know that there must be a path difference of a whole number of wavelengths, $n\lambda$, between them (where *n* is an integer: 1, 2, 3, etc.).
- ■ From Figure 9.16 we can see that the path difference is $d \sin \theta$, so that the *condition for constructive interference* becomes $n\lambda = d \sin \theta$. In other words, constructive interference will occur at angles which have sines of 0, $\frac{\lambda}{d}$, $\frac{2\lambda}{d}$, $\frac{3\lambda}{d}$, etc. (Usually the angles involved with double-slit interference patterns are small, so that $\sin \theta \approx \tan \theta \approx \theta$.)
- ■ In the Young's double-slit experiment, angles cannot be measured directly, so they need to be calculated from distances. Consider Figure 9.17: for the first bright fringe from the centre (the first 'order'), $n = 1$ and $\sin \theta \approx \tan \theta \approx \theta \approx \frac{S}{D}$.
- ■ $n\lambda = d \sin \theta$ can then be re-written as $s = \frac{\lambda D}{d}$, as given above and in Section 4.4.

> **Key concept**
>
> A double-slit interference pattern for monochromatic light consists of a series of light and dark bands (fringes). In the centre of the pattern the bands are approximately the same width and brightness, although the brightness is also affected by diffraction effects (see below).

> **Key concept**
>
> An equation relating wavelength to the spacing of the pattern and the dimensions of the apparatus can be developed by considering the path difference between rays travelling to the same point from the two separate slits.

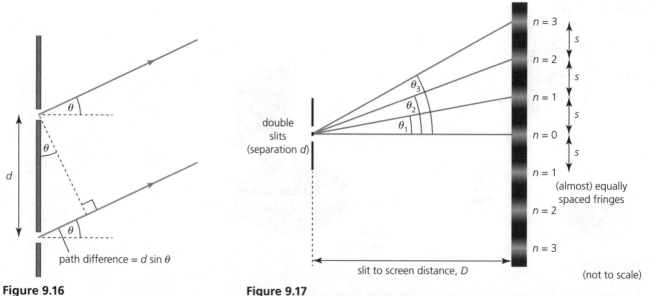

Figure 9.16   **Figure 9.17**

Expert tip

Of course, if the rays in Figure 9.16 are *perfectly* parallel they cannot meet and the waves cannot interfere. However, remember that *d* is very small and the screen may be 2 m away, so that rays meeting on the screen are *very* close to being parallel.

■ Sketching and interpreting intensity graphs of double-slit interference patterns

- The pattern of fringes is commonly shown on a graph representing the relative intensities at different angles. See Figure 9.18, which represents the centre of the pattern.

■ Modulation of two-slit interference pattern by one-slit diffraction effect

- The discussion so far of double-slit interference has ignored one important factor. We have treated both slits as if they were each a *single source of waves*, whereas earlier in this chapter we explained the diffraction patterns from single slits by considering that each slit acts as a source of *many wavelets*.
- We can combine these two effects by simply saying that the double-slit interference pattern is *modulated* by the single-slit diffraction effect. This is shown in Figure 9.19.

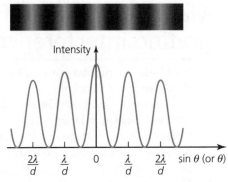

Figure 9.18

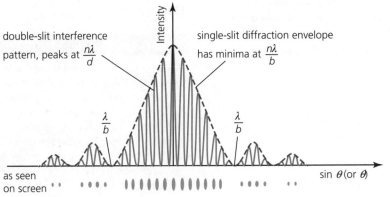

Figure 9.19

> ### Key concept
> The intensity peaks of double-slit interference patterns vary beneath the 'envelope' of single-slit diffraction. The overall effect is that some interference peaks are significantly reduced in intensity, or may even be missing.

QUESTIONS TO CHECK UNDERSTANDING

12 In an experiment such as that shown in Figure 9.14, a student measured the distance between the centre of the pattern and the centre of the sixth bright fringe to be 5.8 cm. If the separation of the slits was 0.085 mm and the screen was 1.66 m from the slits, determine the wavelength of the light used in the experiment.

13 Monochromatic laser light of wavelength 628 nm is incident upon two parallel slits which have a separation of 0.17 mm.

   a Explain the meaning of *monochromatic*.

   b Determine the angles at which the first three peaks from the centre of the interference pattern occur.

14 Figure 9.20 shows the intensity variations seen on a screen during a double slits experiment using laser light of wavelength 5.1×10^{-7} m. The screen was 3.4 m from the slits.

   a Explain why the third order (peak) is missing from the pattern.

   b Determine: **i** the separation of the slits, **ii** the width of each slit.

15 Explain why the pattern seen on a screen when white light passes through double slits (Figure 9.15b) is coloured.

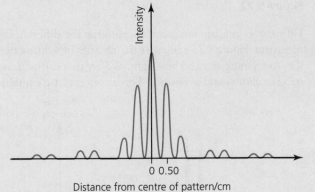

Figure 9.20

Multiple slit and diffraction grating interference patterns

- Double-slit interference patterns are usually blurred and of low intensity, but if the number of identical slits is increased (keeping the same spacing), the patterns for **multiple slits** become sharper (the peaks are narrower) and more intense, as shown in Figure 9.21.
- **Diffraction gratings** have a very large number of parallel slits (*lines*) very close together. They are used to disperse light into spectra for analysis (as discussed in Section 7.1).
- The equation $n\lambda = d\sin\theta$ (as introduced for double slits) can be used for *any* number of identical parallel slits, or, most commonly, for the lines on a diffraction grating (where d is the distance between the centres of neighbouring lines).
- Gratings are distinguished from each other by the number of lines they have per millimetre. For example, a grating with 800 lines mm^{-1} has its lines separated by, $d = 1.25 \times 10^{-6}$ m. If this grating was used with monochromatic light of wavelength 5.0×10^{-7} m, possible values of n could only be 0, 1 and 2 (because $\sin\theta$ has a maximum value of 1). We say that there would be two *diffraction orders*, plus the central order. See Figure 9.22. The angles at which these orders occur can be determined from the highlighted equation.

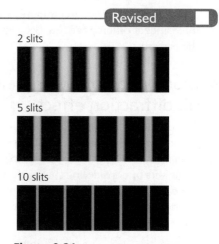

2 slits

5 slits

10 slits

Figure 9.21

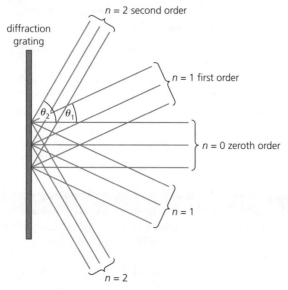

diffraction grating

$n = 2$ second order

$n = 1$ first order

θ_2 θ_1

$n = 0$ zeroth order

$n = 1$

$n = 2$

Figure 9.22

- Diffraction gratings are good at separating the different lines of a line spectrum. Figure 9.23 compares the abilities of a diffraction grating and double slits to separate red and blue light. We say that diffraction gratings produce good *resolution* and *resolvance*. These concepts are explained in Section 9.4.

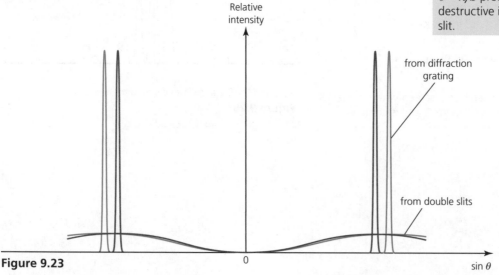

Relative intensity

from diffraction grating

from double slits

Figure 9.23

0

$\sin\theta$

Solving problems involving the diffraction grating equation

- The equation $n\lambda = d\sin\theta$ is commonly used to determine unknown wavelengths using measurements taken from diffraction grating experiments. Measurements are usually made on lines in the first order spectrum because spectra in other orders overlap with each other.

Thin film interference

Revised ☐

- When light is incident upon a *transparent* medium, some light will be reflected off the top surface and some light will be transmitted into the medium. When the transmitted light reaches another surface it may be reflected back so that the waves reflected off two different surfaces may interfere. However, this will only happen if the medium is very thin.
- Commonly seen examples of thin film interference include the effects produced by oil films on water, and soap bubbles. See Figure 9.24.
- Figure 9.25 uses rays to represent the simplified situation in which light waves are only incident *normally* (perpendicularly) on a parallel-sided film. In order to show the reflection clearly, the incident ray has been drawn at a small angle to the perpendicular.

> **Key concept**
>
> If the medium is a transparent and very thin layer (a *film*), the light waves reflected from the two surfaces (interfaces) will be *coherent*, so that interference effects can occur between them.

Describing conditions necessary for constructive and destructive interference from thin films, including phase change at interface and effect of refractive index

- We know that constructive interference occurs when the path difference equals a whole number of wavelengths, and we can see clearly from Figure 9.25 that one wave has travelled a distance of $2d$ more than the other (the total distance travelled in the film).
- But, apart from the actual distances involved, there are two other factors to consider:
  i When light is reflected from the boundary with a medium with a greater refractive index (where it would travel slower) it undergoes a **phase change** of π, which is equivalent to adding a path difference of $\frac{\lambda}{2}$.
  ii If a medium is not air, the wavelength of the light in air must be divided by n (refractive index) in order to determine the wavelength in that medium. For example, if light of wavelength 5.2×10^{-7} m travelling in air enters into water of refractive index 1.3, the wavelength in water will be 5.2×10^{-7})/1.3 = 4.0×10^{-7} m.
- Taking these two factors into account, the condition for *constructive interference* for light incident normally becomes $2dn = (m + \frac{1}{2})\lambda$, where m is an integer. (Note that, to avoid confusion with the symbol for refractive index, this is different from previous equations, in which n was used to represent an integer.)
- The condition for *destructive interference* is $2dn = m\lambda$.
- These two equations apply to the common situations where there are phase changes at one boundary, but not the other. (If the phase changes at both boundaries are the same, the conditions for constructive and destructive interference are reversed).

Figure 9.24

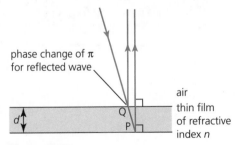

phase change of π for reflected wave

air
thin film of refractive index n

Figure 9.25

- Thin coatings on lenses, solar panels and solar cells reduce the amount of light reflected and increase the percentage of light transmitted. The coating needs to be $\frac{\lambda}{4}$ thick (wavelength in the coating).
- The interference of monochromatic light off the top and bottom of oil films can be used to measure their thicknesses.
- Varying coloured effects are seen when oil or soap films are observed in white light (see Figure 9.24). Figure 9.26 helps to explain these observations: it shows that the path difference between rays reflected from the two surfaces depends on the angle of incidence, so that constructive interference occurs for different colours at different angles. The effects seen will also vary with the thickness of the film.

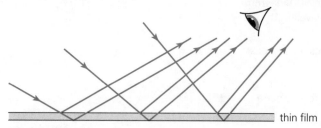

thin film

Figure 9.26

Solving problems involving interference from thin films

- Numerical examples in the IB Physics course will be restricted to normal incidence on parallel-sided films, so that the two highlighted equations above can be used in all calculations where there is a phase change at the top surface (only).

QUESTIONS TO CHECK UNDERSTANDING

19 Light which has wavelength of 405 nm when passing through water of refractive index 1.33 is then transmitted into glass of refractive index 1.52, what will its wavelength become?

20 Does light reflected from an air/water interface undergo a phase change? If so, what is the magnitude of that phase change?

21 a What is the minimum thickness of an oil film on water (refractive index 1.33) which will produce destructive interference for light of wavelength 628 nm? (Refractive index of oil = 1.46.)

　 b Estimate how many molecules thick the film would be.

　 c Suggest how this effect would change if the oil was on glass (refractive index = 1.53) instead of water (no calculation expected).

　 d Determine one thickness of oil film on water which would produce constructive interference for light of wavelength 476 nm.

22 a Explain why anti-reflection coatings on lenses need to be λ/4 thick.

　 b White light has a wide spectrum of different wavelengths; suggest which wavelength would be included in a calculation to determine the thickness need for a lens coating.

■ Curiosity

The fascinating phenomenon of some natural plant and animal surfaces changing apparent colour depending on how they are viewed is called *iridescence*. Of course, scientists were just as curious as everyone else about why these effects occurred. It is usually an interference effect. Natural curiosity is always an important motivation for scientists, and it contributed to them adapting the theory of interference to explain the observations.

■ Serendipity

The first production of thin films was unintentional (Fraunhofer), but it has led on to many useful applications. When an accidental or unplanned event results in good fortune it is called *serendipity*.

9.4 Resolution

Revised ☐

Essential idea: Resolution places an absolute limit on the extent to which an optical or other system can separate images of objects.

- Image resolution should not be confused with *resolution* of vectors into components.
- The ability to resolve objects depends principally on (i) how close together they are, and how far away they are, (ii) the wavelength of the radiation involved and (iii) the properties of the image forming system. In this section we will consider particularly how the *diffraction* of waves entering the system affects resolution.
- Resolution is also affected to some extent by the nature of any material between the objects and the detector. The resolution of binoculars, for example, will be affected by the clarity of the intermediate air.
- Astronomers can get better resolution of images from space by placing telescopes above the Earth's atmosphere.
- When we detect an image (of a point object) through an aperture, the pupil of an eye for example, we are actually receiving a *single-slit diffraction pattern* (see Section 9.2) with most of the energy concentrated in the central maximum.
- The resolution of the system depends on the amount of diffraction that occurs at the aperture where the light (or other kind of wave) is received and detected.
- An understanding of resolution begins with an appreciation of *angular separation*, a concept which expresses the separation of objects as an angle, rather than a distance.
- The **angular separation** of two objects is equal to the angle they subtend at the observer, as shown in Figure 9.27. This may also be called the *angle subtended* at the eye by the two objects.

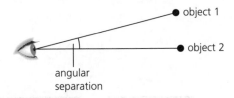

Figure 9.27

Key concept

Good **resolution** describes the ability of an image forming system to see objects which are close together as separate from each other.

There is good resolution if the angular separation between two objects which can be resolved is small.

The ability to resolve two objects as separate depends on how their diffraction patterns overlap after they enter the receiver.

Expert tips

The quality of the lenses, or mirrors, used to focus images, and the separation of the individual sensors at the place where the images are formed will also affect image quality and resolution. The *sensors* may be, for example, the cells on the retina at the back of an eye, or the pixels in a camera.

When digital images are transferred and displayed electronically, for example on the screen of a smart phone or television, we are often concerned with the *display resolution* of the device. But our concern in this section is the original resolution of an image rather than how it is displayed after that.

The resolution of simple monochromatic two-source systems

- To begin an understanding of resolution we keep things simple by considering the images formed from only *two objects which act as point sources*, both of which are the same distance from the detector and both of which emit *monochromatic radiation of the same wavelength and intensity*.

- When we observe these two objects, we will receive two identical, single slit diffraction patterns which may overlap each other if the objects are close together. It is the ability to see the two central maxima as separate from each other that affects the resolution.

- The three diagrams in Figure 9.28 each show the diffraction patterns from two identical objects. In Figure 9.28a the central maxima are well separated and the objects are easily resolvable. In Figure 9.28b the central maxima overlap and their intensities will combine, but a drop in intensity between them will be just detectable, so that they are *just* resolvable. In Figure 9.28c the central maxima are so close that they cannot be distinguished and the sources cannot be resolved.

a Two sources easily resolved

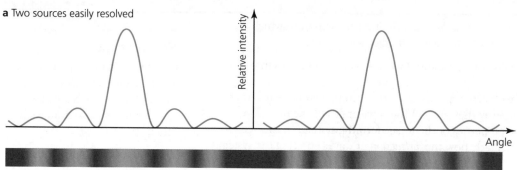

b Two sources *just* resolved

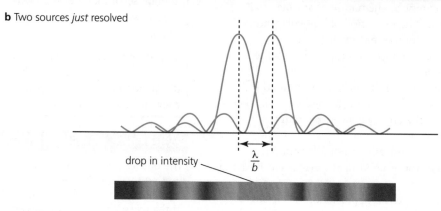

drop in intensity

$\frac{\lambda}{b}$

c Two sources not resolved

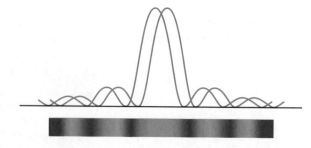

Figure 9.28

- The two patterns shown in Figure 9.28b have an angular separation of $\frac{\lambda}{b}$ and this is taken to be an indication of the limit of resolution. Figure 9.29 represents images of two point sources that are clearly resolved (Figure 9.29a) and which can *just* be resolved (Figure 9.29b).

a  **b**

Figure 9.29

- Rayleigh's criterion is not a physical law, it is just a guide derived from an idealized situation.
- In other words, for light passing through a narrow slit, the criterion suggests that the images of two sources can be resolved if they have an angular separation of $\theta = \frac{\lambda}{b}$ *or greater.*
- Circular apertures are much more common in imaging systems than rectangular slits and their resolution is poorer (for the same width), so that the criterion becomes: the images of two objects detected through a circular aperture can be resolved if they have an angular separation of $\theta = 1.22\dfrac{\lambda}{b}$ or greater.
- The latest scientific research often involves obtaining improved images of objects that may be very small (atomic scale) or very large (astronomical scale). Rayleigh's criterion shows us that two ways of improving resolution are by using larger apertures and/or smaller wavelengths.

Solving problems involving the Rayleigh criterion for light emitted by two sources diffracted at a single slit

QUESTIONS TO CHECK UNDERSTANDING

23 A student looks at two small, identical LEDs separated by 2.35 cm. Both of the lights are 3.50 m from the student.

   a What is the angular separation of the lights?

   b The LEDs emit blue light of wavelength 4.3×10^{-7} m and the student's pupil has a diameter of 2.8 mm. Determine the minimum angle of resolution of her eye.

   c Can she resolve the two lights as separate?

   d Explain how the resolution will change if the LEDs are changed so that they emit red light.

24 a Two similar stars are both 136 000 light years from Earth and they are 1.2 light years apart from each other (a light year is an astronomical unit of distance). Estimate the theoretical minimum aperture width for an optical telescope that can resolve these stars.

   b Suggest why a telescope of this size may in fact be unable to detect the stars.

25 Make a copy of Figure 9.28b on graph paper. On the same axes, use the principle of superposition to sketch a graph which shows the combined intensity of the two patterns.

The size of a diffracting aperture

- For a given wavelength, resolution can be improved by increasing the size of the receiving aperture of cameras, telescopes and microscopes. But only if the quality of the imaging system is unaffected by the increase in size.
- Radio astronomy (see Figure 9.30) must use the relatively long (radio) wavelengths that are emitted from the astronomical sources that they are observing, but their resolution can be improved by making the receiving 'dish' with as large a diameter as possible.

Figure 9.30

Expert tip

Astronomers can improve the resolution of radio telescopes by combining the signals received from an *array* of separate telescopes arranged in a regular pattern. The signals are made to interfere electronically in a way which is similar in principle to the way in which multiple slits improve optical resolution (Section 9.3).

The effect of wavelength on resolution

- Figure 9.31 shows an infrared picture of a cat (the different colours represent different temperatures). The resolution is poor because the wavelengths of infrared radiation are much greater than the wavelengths of visible light with which we would normally see cats.
- If it is possible to do so, any resolution could be improved by using smaller wavelengths. An example would be the use of a blue filter with a microscope to absorb all but the shorter wavelengths in white light. (Using a limited range of wavelengths also improves resolution.) Electron microscopes have greater resolution than optical microscopes because they use electron waves which have very small wavelengths.

Figure 9.31

Resolvance of diffraction gratings

Revised

- We have seen in Section 9.3 that diffraction gratings can be used to analyse spectra by dispersing light from a single source into separate intensity peaks for different wavelengths. See Figure 9.23. The ability to detect separate wavelengths is another kind of resolution.
- Suppose that a diffraction grating is able to resolve light of two similar wavelengths, separated by a difference of $\Delta\lambda$. Resolvance is defined as $R = \dfrac{\lambda}{\Delta\lambda}$ (it has no units because it is a ratio). The higher the value of resolvance, the better the resolution.
- Higher resolution is achieved by the light passing through a larger number of lines (slits), N. Resolvance also improves the higher the diffraction order, m.
- The resolvance of a particular situation can be calculated as follows: $R = mN$.
- Comparing the last two highlighted equations, we see that $\Delta\lambda = \dfrac{\lambda}{mN}$, which enables us to calculate the minimum wavelength difference that can be resolved in any particular situation. If this equation is used to check if two wavelengths can be resolved, the λ used in the equation may be either of the two (or their average).

> **Key concepts**
>
> The ability of a grating to separate wavelengths is called its **resolvance** (or *resolving power*), R.
>
> If light is incident on a greater number of lines/slits in a grating, the resolvance is improved.

QUESTIONS TO CHECK UNDERSTANDING

26 An optical instrument produces an angular resolution of 5.0×10^{-4} rad when looking at two objects in red light. Estimate the resolution that would be achieved by using blue light instead.

27 Greater resolution can be achieved by using image forming instruments with larger apertures.

  a Suggest one other advantage of using larger apertures.

  b Suggest one disadvantage of instruments with larger apertures.

28 a Explain why radio telescopes are much larger than optical telescopes.

  b Estimate the resolution of the radio telescope shown in Figure 9.30 when it is receiving 21 cm radio waves emitted from hydrogen in a distant galaxy.

29 a When light is incident upon 25 slits in a diffraction grating, determine if wavelengths of 490 nm and 500 nm can be resolved in the first, second and third diffraction orders.

  b If the diffraction grating had 300 lines mm⁻¹, what was the cross-sectional area of the incident circular light beam?

NATURE OF SCIENCE

Improved technology

Scientific advances are often preceded by improvements in equipment and instrumentation. This is particularly true in image formation technology, where the latest developments are producing some resolutions which are better than the simplified Rayleigh limit.

9.5 Doppler effect

Essential idea: The Doppler effect describes the phenomenon of wavelength/frequency shift when relative motion occurs.

The Doppler effect for sound waves

- When our ears hear a sound we usually assume that the frequency that we detect ('observe') is the same as the frequency emitted by the source. But if there is relative motion between a source of sound and the detector/observer (an ear, for example) the two frequencies will be different. This is known as the **Doppler effect** (sometimes called the *Doppler shift*).

■ ### Sketching and interpreting the Doppler effect when there is relative motion between source and observer

- Figure 9.32 shows circular wavefronts between a source and a detector in three different situations.

- In Figure 9.32a there is no motion, so that the detector receives the same frequency as that emitted by the source. In Figure 9.32b the detector is moving directly towards a stationary source and meets more wavefronts every second than it would if it was not moving. This results in detecting an increased, but *constant* frequency. In Figure 9.32c a sound emitting source is moving towards a stationary detector. This results in the wavefronts being squashed closer together, and the detector receives an increased, but *constant* frequency.
- Similar diagrams can be drawn to represent situations where the distance between a source and detector is increasing.

a Source and detector both stationary

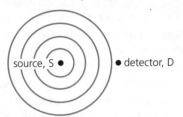

b Detector moving towards stationary source

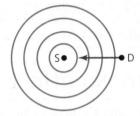

c Source moving towards stationary detector

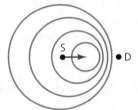

Figure 9.32

Common mistakes

When a source of sound travels in a straight line (almost) directly towards and then away from an observer, the frequency changes suddenly from a higher to a lower value as the source passes the observer. But if the relative motion is not *directly* between the source and detector (Figure 9.33), the frequency received will not be constant. Figure 9.33b is a common situation: the *changing* frequency received from a vehicle which is moving quickly past an observer.

a

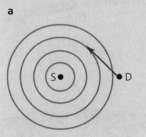

b

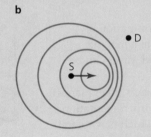

Figure 9.33

■ ### Solving problems involving the change in frequency or wavelength observed due to the Doppler effect to determine the velocity of the source/observer

- *If a source of sound is moving* directly towards or away from a stationary observer/detector with a speed u_s, the detected frequency, f', can be determined from the following equation:

$$f' = f\left(\frac{v}{v \pm u_s}\right)$$

where the emitted wave frequency is f and the wave speed is v. A received frequency is greater when the separation is decreasing (subtract u_s), and less when the separation is increasing (add u_s).

Key concept

An observer receives a different sound frequency from that which was emitted if the source or observer is moving. The magnitude of the shift in frequency depends on how the speed of movement compares to the speed of sound.

■ *If an observer/detector is moving* directly towards or away from a stationary source of sound with a speed u_0, the detected frequency, f', can be determined from the following equation:

$$f' = f\left(\frac{v \pm u_0}{v}\right)$$

A received frequency is greater when the separation is decreasing (add u_0), and less when the separation is increasing (subtract u_0).

QUESTIONS TO CHECK UNDERSTANDING

30 Draw a diagram to show the wavefronts between an observer and a source of sound which is moving directly away.

31 A train is travelling on a straight track towards a station when it emits a sound of frequency 220 Hz. An observer at the station records a change in frequency of 12 Hz from the frequency emitted.

   a   What frequency sound was received at the station?

   b   Assuming the speed of sound was 335 m s⁻¹, what was the speed of the train when it emitted the sound?

   c   If, after passing through the station the speed of the train increased to 33 m s⁻¹, what frequency would then be recorded at the station?

32 Sound of wavelength 90 cm is emitted from a stationary source. What wavelength will be received by an observer moving at 14 m s⁻¹ towards the source? Assume that the speed of sound is 340 m s⁻¹.

The Doppler effect for light waves

■ The Doppler shift also occurs with *electromagnetic waves*, for example, light.

■ For electromagnetic waves we do *not* use the equations highlighted above. Instead the *change* in frequency or wavelength caused by the Doppler effect can be calculated from: $\frac{\Delta f}{f} = \frac{\Delta \lambda}{\lambda} \approx \frac{v}{c}$. This is an approximation and it can only be used accurately if the relative speed, v, between source and detector is much less than the speed of electromagnetic radiation, c (which is usually true).

Describing situations where the Doppler effect can be utilized

■ An important application of the Doppler effect is in the determination of speeds: if waves of a known speed and frequency are directed at a moving object, the *change* of frequency (or wavelength) of the reflected waves can be used to calculate the speed of that object.

■ *Radar* is a system for detecting the location and velocity of objects (likes planes and cars) which are often significant distances away from the instrumentation. Radio (microwave) pulses are sent from a transmitting aerial which may rotate (see Figure 9.34), they reflect off objects and some of that reflected energy is received back at the aerial. The delay between emitted and received pulses can be used to calculate the distance to the object (assuming that the speed of the waves is known).

■ *Doppler radar* incorporates the use of the Doppler effect to determine the velocity of a moving object. It is also widely used to track weather systems.

■ Because of an effect similar to the Doppler effect, the frequencies of radiation from distant galaxies (or stars) which is received on Earth are slightly *lower* than the frequencies emitted by the galaxies. The emitted frequencies are well known from the *line spectra* (Chapter 7) of the same elements on Earth, as shown in Figure 9.35.

Figure 9.34

■ The shift of a spectrum to *lower* frequencies is called a **red shift**, it is evidence that the distance between galaxies is increasing: the universe is expanding.

line spectrum from element on Earth

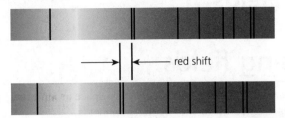

red shift

line spectrum from element in distant galaxy

Figure 9.35

■ The Doppler effect is also used with ultrasound in hospital to measure the velocity of blood in arteries and veins. See Figure 9.36.

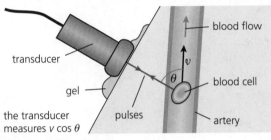

transducer

gel

the transducer measures $v \cos \theta$

pulses

blood flow

blood cell

artery

Figure 9.36

QUESTIONS TO CHECK UNDERSTANDING

33 If the radar used to track aircraft has a wavelength of 20 cm, what change of wavelength will be detected after the microwaves have reflected off a plane travelling at a speed of 260 m s⁻¹ towards the airport?

34 A line in the hydrogen line spectrum has a frequency of 6.563×10^{14} Hz. When the same spectrum is emitted from a distant galaxy and viewed on Earth, the frequency has changed to 6.541×10^{14} Hz.

a What is this change of frequency commonly called?

b Determine the speed of the galaxy.

c In which direction is the galaxy moving?

d Use of the equation $\frac{\Delta f}{f} \approx \frac{v}{c}$ involves an assumption. Is that assumption valid in this question?

35 The medical examination shown in Figure 9.36 involves *ultrasound*.

a What is ultrasound?

b Suggest what properties of ultrasound make it useful for this examination.

NATURE OF SCIENCE

■ **Technology**

Doppler first identified the effect that bears his name more than 170 years ago. The effect now has many useful applications which would not have been possible until technological advances enabled the accurate measurement of relatively small changes of frequency.

10 Fields

10.1 Describing fields

Essential idea: Electric charges and masses each influence the space around them and that influence can be represented through the concept of fields.

■ Electric forces between charges and gravitational forces between masses have been discussed in Chapters 5 and 6. These forces can act across free space and the concept of field was introduced to help explain such 'actions at a distance'.

■ Reminder from Chapters 5 and 6: $F_E = \frac{kq_1q_2}{r^2}$ and $F_G = \frac{Gm_1m_2}{r^2}$ (m_1m_2 is sometimes written as Mm).

■ Reminder from Chapters 5 and 6: the strength of an electric field, or the strength of a gravitational field at any point can be found by determining the force exerted on a small (positive) test charge, or a small test mass, placed at that point: $E = \frac{F}{q}$, $g = \frac{F}{m}$. (A *test charge* or *test mass* is considered to have no effect on the field that it is determining.)

Gravitational fields

■ There are many similarities in the theories and mathematics of gravitational and electric fields. An understanding of one greatly helps the understanding of the other. We will begin with gravitational fields because they are generally simpler and more familiar to us.

■ Representing sources of mass, lines of gravitational force and field patterns using an appropriate symbolism

■ Although they are clearly not visible, it is very helpful to be able to represent gravitational fields on paper or screens. There are two inter-related ways of doing this: using *lines of equal potential* (see later) and/or *field lines* which show the directions of forces:

■ Field lines

■ Field lines can never cross each other and, in any given diagram, the field strength is greatest where the lines are closest together.

■ It should be appreciated that field lines on paper are a two-dimensional representation of a three-dimensional field.

■ In this course we will restrict our discussion of gravitational field lines to (i) radial fields around spherical masses (like planets, moons and stars) and (ii) the uniform fields close to the surfaces of such masses.

■ Figure 10.1 shows the *radial field lines* pointing inwards towards a (theoretical) *point mass*, and the field around an equal mass, but which is larger and *spherical* (for example, a planet, moon or star). Note that the field at any place *outside* of the surface of a sphere (of uniform density) is the same as it would be for a point of the same mass.

> **Key concept**
>
> **Gravitational field lines** show the direction of gravitational force on masses at points within the field.

Figure 10.1

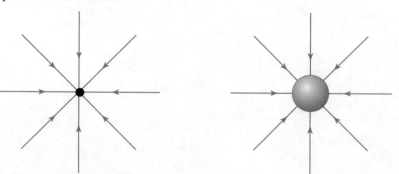

- Figure 10.2 shows the uniform field that is assumed to exist above a small section of the surface of a large planet. The field is in fact radial, but is effectively uniform on this relatively small scale.

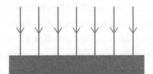

Figure 10.2

Gravitational potential energy

- When masses are moved in gravitational fields energy is transferred (work is done), unless the movement is perpendicular to the direction of the force (and field). We describe these as *changes* in *gravitational potential energy*.

- In Chapter 2 we used the equation $\Delta E_P = mg\Delta h$ to calculate the *change* in gravitational potential energy, ΔE_P, when a mass, m, is moved up or down a distance Δh. But it should be stressed again that (i) this equation can only be used in a *uniform* gravitational field, g, such as that close to the Earth's surface, and (ii) the use of this equation does not determine *total* values of gravitational potential energy.

- To move masses apart, work has to be done against the gravitational forces attracting them together. That is, energy must be transferred *to* the system.

- Conversely, when gravitational forces pull masses closer together, energy will be transferred *from* gravitational potential energy to some other form (usually kinetic energy).

- More generally, we consider that gravitational potential energy is stored in any system of masses because, at some time in the past, work was done when the masses moved to their present positions.

- Before we can think about how to calculate the *total* gravitational potential energy of any system (rather than changes), we need to be clear about an agreed *zero* for gravitational energy.

- In practice, we will nearly always be discussing the movement of relatively small masses to, or from, much larger masses (like the Earth or another planet), so that we commonly only refer to transferring energy to, or from, the smaller mass (rather than a system of two or more masses).

- To move a mass from the Earth, for example, to a very long way away (infinity) requires that we transfer a lot of energy to it, and so increase its gravitational energy. Because at infinity the gravitational potential energy is zero, it means that *all* gravitational potential energies must be negative.

- For example, when any mass moves closer to a planet it will lose gravitational potential energy, so that the magnitude of its potential energy has a larger negative value. The gravitational potential energy of a mass moving away from a planet increases, becoming a smaller negative value. See Figure 10.3, which represents the theoretical situation of a mass projected to infinity from the Earth's surface.

- But note that *changes* of gravitational potential energy may be positive (increasing separation) or negative (decreasing separation). This is equally true for a spacecraft, or a ball thrown in the air.

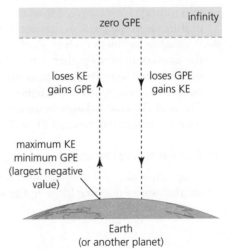

Figure 10.3

- For example, we might refer to a mass (of 12kg) on the surface of the Earth as having gravitational potential energy of, say, -7.51×10^8 J. This simply means that we would need to transfer $+7.51 \times 10^8$ J to the mass in order to move it a very long way from Earth (to infinity). This ignores any factors that might dissipate energy (like air resistance in the atmosphere).

Common mistake

It may be thought that to move a mass to infinity requires an infinite amount of energy, but that is not true. Because gravitational forces follow an inverse square law with distance, in theory they never reduce to zero, but the forces become vanishingly small with the enormous distances involved in astronomy.

QUESTIONS TO CHECK UNDERSTANDING

1 Explain why the equation $\Delta E_P = mg\Delta h$ cannot be used to determine the increase in gravitational potential energy when a satellite is put into orbit around the Earth.

2 A satellite in orbit around the Earth has gravitational potential energy $-E_P$.

   a Explain why this energy is negative.

   b If the satellite is moved to a lower orbit does its gravitational potential energy increase or decrease?

   c How does the magnitude of the gravitational potential energy change?

3 The gravitational force on a mass of 1.0 kg on the Earth's surface is about 10 N. Calculate the gravitational pull from Earth on 1.0 kg which is 6.0×10^{11} m away (approximately equivalent to the distance from Earth to the planet Jupiter). The Earth's radius is 6.4×10^6 m.

■ Gravitational potential

- We have seen that, in a particular situation, we can calculate the *force* involved when a certain mass is moved in certain gravitational field, but the concept of *gravitational field strength* (force per unit mass) enables us to generalize to all masses. In the same way, we can calculate the *gravitational potential energy* involved when a certain mass is moved in a gravitational field, but the concept of *gravitational potential* (gravitational potential energy per unit mass) enables us to generalize to all masses.

- As with gravitational potential energy, gravitational potentials are always negative (for the same reason: infinity has zero potential).

- For example, the gravitational potential on the Earth's surface is -6.26×10^7 J kg^{-1}. Using this information we can simply multiply by the mass of any object to determine the amount of energy that would be needed to transfer it to infinity. For example, a mass of 12 kg would need 7.51×10^8 J to reach infinity (as in the previous sub-section).

- Because we are more likely to want to discuss movements of masses close to the Earth or in orbits (rather than sending a spacecraft a very great distance from Earth), the *gravitational potential difference* between two points is usually a more useful concept than gravitational potential.

- The work, W, that is done when a mass, m, moves between two places which have a potential difference of ΔV_g is determined from $W = m\Delta V_g$.

■ Mapping fields using potential

- *Equipotential lines* (or in three dimensions, *equipotential surfaces*) provide another way of mapping (drawing) gravitational fields.

Key concepts

Gravitational potential, V_g, at a point is defined as the work per unit mass (kg) needed to bring a small test mass from infinity to that point (unit: J kg^{-1}).

Gravitational potential difference, ΔV_g, is defined as work per unit mass (kg) needed to move a small test mass between the two points (unit: J kg^{-1}).

■ Equipotential surfaces

- ■ Contour lines on a map are examples of gravitational equipotential lines.
- ■ Equipotential lines do not have direction and, like field lines, they can never cross each other.
- ■ By definition, there is no *potential difference* between points on the same equipotential line or surface, and therefore no net work is done when a mass moves between such points.

> **Key concept**
>
> **Equipotential lines** and **equipotential surfaces** connect places which have the same potential. Equipotential lines are always perpendicular to field lines.

■ Describing the connection between equipotential surfaces and field lines

- ■ Figure 10.4 shows (in black) some equipotentials drawn around a spherical planet. The field lines, which are perpendicular to the equipotential lines, are also shown in red. Note that the numerical difference in potential between adjacent lines is kept constant. The spacing of the equipotential lines increases with distance from the mass because the field strength is decreasing, so that greater distances are required to transfer the same amount of energy.

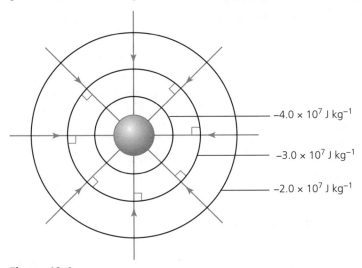

-4.0×10^7 J kg^{-1}

-3.0×10^7 J kg^{-1}

-2.0×10^7 J kg^{-1}

Figure 10.4

- ■ At greater distances from the planet, less energy would be needed to transfer a mass to infinity, so that the gravitational potential (and potential energy) have increased, to become smaller negative values. The field lines point from higher potential to lower potential.
- ■ Figure 10.5 shows the equipotentials around two equal, large masses relatively close together.

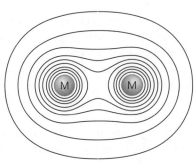

Figure 10.5

QUESTIONS TO CHECK UNDERSTANDING

4 Consider Figure 10.4.

 a How much gravitational potential energy would need to be transferred to a satellite of mass 2000 kg in orbit around the planet to move it between the heights shown by the inner equipotential and the outer equipotential?

 b A satellite in a circular orbit moves along an equipotential surface. Explain why it does not need to be powered to follow such a path.

5 Use the information shown in Figure 10.5 to sketch the gravitational field lines around two equal masses. Mark a point where the field is zero (the resultant force on a mass at that point would be zero).

6 What is the gravitational potential difference between sea level and the top of Mount Everest (8848 m)? What assumption did you make?

7 a What value would you give to an equipotential surface passing through the room where you are reading this question?

 b Describe the orientation of that equipotential.

8 The gravitational potential on the surface of a planet is -4.96×10^7 J kg^{-1}.

 a Determine the minimum amount of work that has to be done to move a mass of 847 kg completely away from the gravitational attraction of the planet.

 b Suggest reasons why, in practice, much more energy will be involved in this process.

Electrostatic fields

- In this section we will discuss the electric fields around charges which are stationary (**electrostatic fields**).
- The concepts of (gravitational) potential energy and potential discussed in the previous section can be applied by *analogy* to electrostatic fields. For the sake of clarity they are repeated below (as applied to electric fields), but electric fields are also more complicated because there are two kinds of charge (positive and negative) but only one kind of mass: electric forces can be attractive or repulsive, but gravitational forces are only attractive.

■ Representing sources of charge, lines of electric force, and field patterns using an appropriate symbolism

- Although they are invisible, it is very helpful to be able to represent electric fields on paper or screens. There are two inter-related ways of doing this: using *lines of equal potential* (see later) and/or *field lines* which show the directions of forces:

■ Field lines

- The representation of electric fields by field lines has been covered in Section 5.1.

> **Key concept**
>
> **Electric field lines** show the direction of electric forces on *positive* (test) charges.

■ Electric potential energy

- When charges are moved in electric fields energy is transferred (work is done), unless the movement is perpendicular to the direction of the force (and field). We describe these as *changes* in electric potential energy.
- To move *opposite charges* apart, work has to be done against the electric forces attracting them together. That is, energy must be transferred *to* the system.
- Conversely, when electric forces pull *like charges* closer together, energy will be transferred *from* electric potential energy.
- More generally, we consider that *electric potential energy* is stored in any system of charges because, at some time in the past, work was done when the charges moved to their present positions.
- If we wanted to calculate the *total* electric potential energy of any electrostatic system, we would need to be clear about an agreed *zero* for electric potential energy: a system of static charges has zero electric potential energy if they are so far apart (an *infinite* distance) that there are zero electric forces between them.
- The electric potential energy stored between opposite charges is negative (like gravitational potential energy) because energy would need to be supplied to completely separate them and produce a system with zero energy. The electric potential energy stored between like charges is positive.

> **Key concept**
>
> **Electric potential energy** of a charge at any point, E_P, is the amount of energy needed to bring the charge from infinity to that point.

> **Expert tip**
>
> In electrical circuits the concept of a theoretical zero of electric potential energy at infinity is of little practical use, and connections made to the *ground* establish the *earth* as a realistic zero for electric potential energy and potential.

QUESTIONS TO CHECK UNDERSTANDING

9　Sketch the field lines around
   a　a negatively charged sphere
   b　two negatively charged spheres which are close together, but one sphere has twice the charge of the other.

10 Consider a hydrogen atom. The magnitude of the electric potential energy of the atom is about 2×10^{-18} J.

   a Is this energy positive or negative?

   b Express the energy in electronvolts.

   c How much energy would be needed to separate the electron from the proton?

   d After separation (ionization), what is the total electric potential energy of the system?

11 Two isolated point positive charges are separated by a small distance.

   a Is the electric potential energy stored in this situation positive or negative?

   b Explain the energy changes that will occur if the charges are able to move freely.

12 In a thunder storm a large amount of negative charge builds up on the lower surface of a cloud. This repels some negative charge in the ground below, leaving it positively charged. Sketch the general shape of the electric field pattern between the cloud and the ground.

■ Electric potential

- Because we are unlikely to want to discuss movements of charge to infinity, the electric potential *difference* between two points is usually a more useful concept than electric potential.
- The work, W, that is done when a charge, q, moves between two places which have a potential difference of ΔV_e is determined from $\mathbf{W = q\Delta V_e}$.
- The concept of transferring energy to, or from, charges is central to an understanding of electric circuits. *Potential difference (p.d.) in circuits* was discussed at length in Chapter 5, where the unit for p.d., the volt, V, was introduced as the widely used alternative to JC^{-1} (p.d. is commonly called *voltage*).

> ### Key concepts
>
> **Electric potential**, V_e, is the work per unit charge (C) that would need to be done to bring a small test positive charge from infinity to that point. Unit: JC^{-1} (or volt, V).
>
> **Electric potential difference** between two points, ΔV_e, is the work per unit charge (C) that would need to be done to move a small test positive charge between the two points. Unit: JC^{-1} (or volt, V).

> ### Common mistake
>
> The basic and very useful concepts of *potential* and *potential difference* cause difficulty for many students. This is partly because of the names themselves and the name of the unit given to electric potential: the volt (there is no name given to the gravitational equivalent). It may reduce confusion by remembering that potential difference is simply energy/charge transfered and the volt is just another way of writing joules per coulomb.

■ Mapping fields using potential

- *Equipotential lines* (or in three dimensions, *equipotential surfaces*) provide another way of mapping (drawing) electrostatic fields.

■ Equipotential surfaces

- Equipotential lines do not have direction and, like field lines, they can never cross each other.
- By definition, there is no *potential difference* between points on the same equipotential line or surface, and therefore no net work is done when a charge moves between such points.

> ### Key concept
>
> **Equipotential lines** and **equipotential surfaces** connect places which have the same potential.
>
> Equipotential lines are always perpendicular to field lines.

■ Describing the connection between equipotential surfaces and field lines

- Figure 10.10 shows (in black) some equipotentials drawn around charged spheres. The field lines, which are perpendicular to the equipotential lines, are also shown in red. Note that the potential difference between adjacent equipotential lines is constant. The spacing of the equipotential lines increases with distance from the charges because the field strength is decreasing, so that greater distances are required to transfer the same amount of energy.

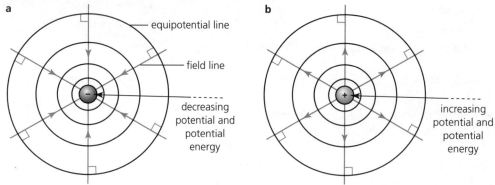

Figure 10.10

- Figures 10.11 and 10.12 show the equipotentials and field lines around pairs of charges (of equal magnitude).

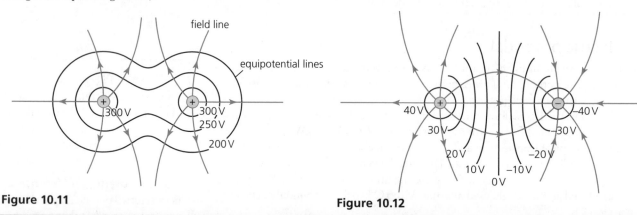

Figure 10.11   **Figure 10.12**

QUESTIONS TO CHECK UNDERSTANDING

13 a How much energy is transferred when a charge of −26 nC moves between places where the potentials are +1.0 kV and +1.5 kV?

b Explain why the answer may be positive or negative.

14 Two parallel metal plates are connected to a p.d. of 10 V. The lower plate is connected to earth to keep it at 0 V. Sketch the shape of four equipotential lines between the plates and a little beyond their edges. Label their values.

15 Figure 10.13 shows the equipotentials near a charged sphere placed near an earthed metal plate.

a Copy the diagram and put values on the two unlabelled lines.

b Is the sphere positively or negatively charged?

c Add lines to represent the shape and direction of the electric field.

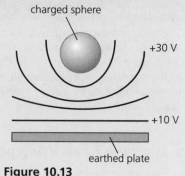

Figure 10.13

NATURE OF SCIENCE

■ Paradigm shift

When the generally accepted way of thinking about a situation completely changes it is called a *paradigm shift*. Many significant advances in science have arisen because a problem has been interpreted in a totally new way, and old theories have been discarded. The rejection of widely-held beliefs and the introduction of such paradigm shifts requires great imagination and it may be difficult for some people to change their ways of thinking (the previous paradigm).

10.2 Fields at work

Revised ☐

Essential idea: Similar approaches can be taken in analysing electrical and gravitational problems.

- Figure 10.14 summarizes the four closely inter-connected concepts which physicists use to describe fields: *force, field strength, potential energy* and *potential*.

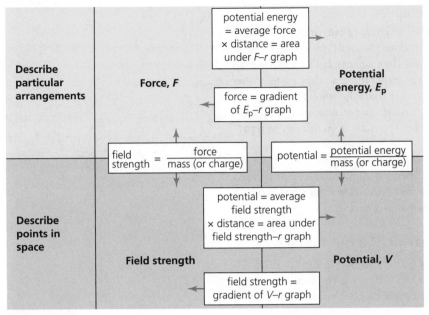

Figure 10.14

■ In Section 10.2 we will go into more detail and develop the mathematics to deal with the most basic type of field: the *radial fields* around point masses or point charges (or spheres). Once again, we will begin with gravitational fields.

Radial gravitational fields

Revised ☐

■ Forces and inverse square law behaviour

■ Consider Figure 10.15, which shows a large spherical mass, M, around which there is a radial gravitational field. From Section 6.1, we know that if another mass, m, is at point P it will experience a gravitational force $F_G = \frac{GMm}{r^2}$ towards M (and M will experience an equal and opposite force). This is Newton's law of gravitation and it is sometimes written in the form $F_G = \frac{Gm_1 m_2}{r^2}$.

■ In this section we are concentrating on the effects of a strong radial gravitational field around a large mass, M (like a planet), on a much smaller mass, m (like a satellite), in such a field.

■ A graphical representation of the *inverse square variation* of gravitational force with distance can be seen in Figure 10.16.

■ *Gravitational field strength, g,* also follows an inverse square law variation with distance around a point mass or sphere.

■ Gravitational field strength, $g = \frac{F_G}{m}$, so that at point P, $g = \frac{GM}{r^2}$.

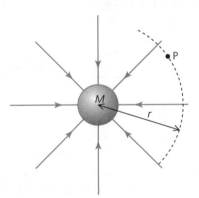

Figure 10.15

■ Solving problems involving forces on masses in radial and uniform fields

QUESTIONS TO CHECK UNDERSTANDING

16 Assuming that the gravitational field on the surface of Mars is uniform ($3.8\,N\,kg^{-1}$), determine the gravitational force acting on a 75 kg astronaut.

17 Determine the gravitational force acting on a satellite of mass 2000 kg orbiting at a height of 550 km above the Earth's surface (radius of Earth = 6.4×10^6 m). The mass of the Earth is 5.97×10^{24} kg.

18 a The mass of the Sun is 3.33×10^5 times greater than the mass of the Earth. Determine the gravitational forces acting on both bodies. They are separated by a distance of 1.50×10^{11} m.

   b Describe the effect of this force on the Earth.

■ Potential energy

■ Gravitational potential energy is stored in a system of two or more masses. We will be considering situations in which a relatively small mass (like a satellite) is moved in the field of a much larger mass (like a planet). The smaller mass will be considered to have no effect on the larger mass, so that we may refer to the potential energy of the small mass (rather than the system).

■ We have seen in Section 10.1 that the *gravitational potential energy*, E_P, of a mass at any point is the work done to bring the mass from infinity to that point.

■ In Section 2.3 we learned that work done can be determined from force multiplied by distance but, if the force varies, an *average* value must be used in the calculation. From Section 2.3, we also know that work done can be determined from the area under a force–distance graph.

■ Determining the potential energy of a point mass

■ Gravitational potential energy, E_P, may be determined from the (shaded) area under a graph like that in Figure 10.16, where r is the distance from the centre of the large mass to the point mass being considered. But note that the gravitational potential energy is always given a negative value, as discussed in Section 10.1.

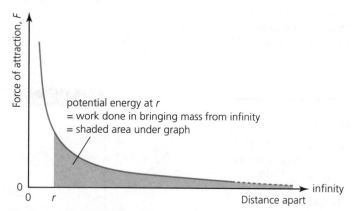

Figure 10.16

■ The gravitational potential energy stored between two spherical masses, M and m, may be determined directly using the equation: $E_P = \dfrac{-GMm}{r}$ (where r is the distance between their centres). There is an inverse relationship between gravitational potential energy and the distance from a point mass (*not* inverse square).

■ Solving problems involving potential energy

QUESTIONS TO CHECK UNDERSTANDING

19 Figure 10.17 shows how the force acting on a mass of 8.5 kg varies with distance from a planet of radius $R = 5.8 \times 10^7$ m.

   a Use the graph to estimate the gravitational potential energy of the mass when on the planet's surface.

   b Determine the gravitational field strength on the planet's surface.

20 Calculate the gravitational potential energy of a 840 kg satellite in an orbit of radius 3600 km around the planet Mars (mass = 6.4×10^{23} kg).

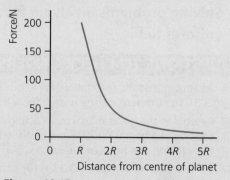

Figure 10.17

Potential

- We know that the *gravitational potential*, V_g, at a point is defined as the work per unit mass that would need to be done to bring a small test mass from infinity to that point. It can be calculated for a mass at any point in a radial field by dividing the gravitational potential energy by the mass,

$$m: V_g = \frac{E_P}{m} = \frac{-GM}{r}.$$

- Figure 10.18 shows the variation of potential with distance from the surface of a planet of radius R. There is an inverse relationship between potential and distance (*not* inverse square).
- Gravitational equipotentials around a spherical mass were shown in Section 10.1, Figure 10.4.

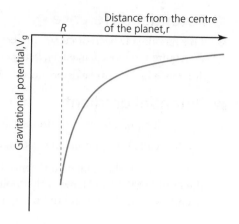

Figure 10.18

Combining field strengths and potentials

- A mass may be in two gravitational fields, for example a spacecraft travelling between the Earth and the Moon is in the fields of both objects.
- Gravitational force and field strength are vector quantities and the combined force or field may be determined by vector addition.
- Gravitational potential energy and potential are scalars, and the combination of two energies or potentials can be determined by simple addition. As an example, Figure 10.20 shows the approximate shape of the variations of potentials (or potential energies) between the Earth and the Moon.

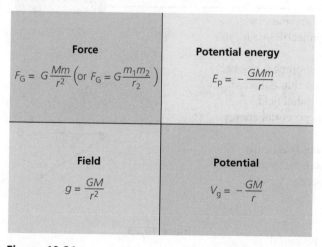

Figure 10.20

- At the position X the magnitude of the combined potentials (or potential energies) is a minimum.

Summary of the four inter-connected radial gravitational field concepts

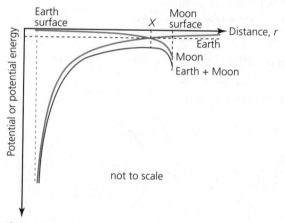

Figure 10.21

> ### Expert tip
>
> A mass which has insufficient kinetic energy to escape from a gravitational field may be described as being in a *gravitational potential well*, such as shown in Figure 10.19. This is a two-dimensional representation of the graph shown in Figure 10.18. More massive planets and stars are represented by deeper wells, from which it is more difficult for other masses to escape.

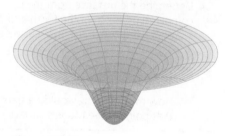

Figure 10.19

Figure 10.21 content:

Force	Potential energy
$F_G = G\frac{Mm}{r^2}$ (or $F_G = G\frac{m_1 m_2}{r_2}$)	$E_p = -\frac{GMm}{r}$
Field	**Potential**
$g = \frac{GM}{r^2}$	$V_g = -\frac{GM}{r}$

■ Potential difference

■ This has already been discussed in Section 10.1. Reminder: the energy (work), W, required to move a mass, m, between points which have a potential difference of ΔV_g can be determined from $W = m\Delta V_g$.

■ Potential gradient

■ The gradient of a potential–distance graph, like that shown in Figure 10.18, is $\frac{\Delta V_g}{\Delta r}$. It is called the *gravitational potential gradient*.

■ In Figure 10.20 the combined potential gradient at X is zero, indicating that the resultant field and force at that point are zero. The gravitational field strengths of the Moon and the Earth are equal and opposite.

> **Key concept**
>
> **Gravitational potential gradient** is equal to the gravitational field strength, g at that point: $g = \frac{-\Delta V_g}{\Delta r}$.

QUESTIONS TO CHECK UNDERSTANDING

21 a Determine the potential at a height of 1000 km above the surface of the Moon. (Mass of Moon is 7.3×10^{22} kg; radius of Moon is 1700 km.)

  b What important assumption did you make?

  c How much energy would need to be transferred to a 50 kg mass to raise it from the Moon's surface to a height of 1000 km?

22 a Plot a graph of the variation of gravitational potential around the Earth with distance from the centre (graph to include distances up to 2.0×10^7 m). (Mass of Earth is 6.0×10^{24} kg; radius of Earth is 6400 km.)

  b Use your *graph* to determine the field strength at a height of 1.0×10^7 m above the surface.

  c Compare your answer to the value calculated using $g = \frac{GM}{r^2}$.

23 Figure 10.22 shows some equipotential lines around the Earth. The potential at the Earth's surface is −63 MJ kg⁻¹.

  a What is the potential difference when moving from the inner line to the outer line?

  b How much gravitational energy must be given to a 500 kg object to raise it from the Earth's surface to the height of the second equipotential?

  c Explain why no net energy is transferred if a satellite moves from A to B.

24 Determine the distance from the centre of the Earth to a point where the combined potential of the Earth and the Sun is at its lowest (gravitational forces are equal and opposite). Sun–Earth distance is 1.5×10^{11} m, mass of Sun = 2.0×10^{30} kg, mass of Earth is 6.0×10^{24} kg.

25 a Calculate the combined gravitational potential of a point which is 2.8×10^8 m from Earth and 3.7×10^7 m from the Moon.

  b How much energy would be transferred to move a 2400 kg spacecraft from that point to the Moon's surface?

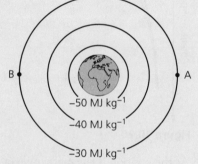

−50 MJ kg⁻¹
−40 MJ kg⁻¹
−30 MJ kg⁻¹

Figure 10.22

Spacecraft in radial gravitational fields

> **Revised** ☐

■ To raise any object away from a planet's surface it has to gain gravitational potential energy. An object may be given kinetic energy (a projectile), or it may be continually powered, like a plane or a rocket.

■ In Chapter 2 we saw that we could calculate the maximum theoretical height, h, reached by a projectile fired vertically from $mgh = \frac{1}{2}mv^2$, but that equation can *only* be used if the field is uniform (g is constant). In the radial field around a planet we need to use the equation for gravitational potential energy in a radial field.

■ Consider Figure 10.23. If a mass, m, at point A is to be projected vertically from the surface of a planet to point B, it needs to be given kinetic energy equal to the difference in gravitational potential energy between A and B:

$$\frac{1}{2}mv^2 = \left(\frac{-GMm}{r_B}\right) - \left(\frac{-GMm}{r_A}\right)$$

■ Alternatively, we can divide through by m, leading to $\frac{1}{2}v^2$ = potential difference, ΔV_g.

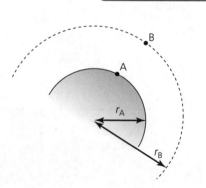

Figure 10.23

- Of course, this equation assumes that there is no significant air resistance.
- Now we will consider the possibility of launching a mass into space with sufficient kinetic energy that it could completely escape a planet's gravitational field and travel to 'infinity'.

Escape speed

- Escape speed does not depend on the magnitude of the mass of the object, or the direction of launch. However, if the planet has an atmosphere (like the Earth), energy would be dissipated because of friction and a much higher speed would be needed (if it was even possible).
- The escape speed, v_{esc}, from a planet (or moon, or star) of mass M and radius r, can be determined by equating the necessary initial kinetic energy to the magnitude of the change in gravitational potential energy between the planet's surface and infinity: $\frac{1}{2}mv_{esc}^2 = \frac{GMm}{r} - 0$, which leads to $v_{esc} = \sqrt{\frac{GM}{r}}$.

Orbital motion and orbital speed

- A *satellite* is the word that we use to describe a moon, planet or vehicle which orbits a much larger mass. Most commonly, the word *satellite* is used to describe artificial satellites that have been placed in orbit around the Earth.
- In Chapter 6 we discussed satellite orbits as examples of circular motion. Equating gravitational force with centripetal force showed that the **orbital speed** required for *any* mass in a circular orbit of radius r around a mass M was given by $v_{orbit} = \sqrt{\frac{GM}{r}}$.
- Only circular orbits are considered in this course. Combining the last equation with $v = \frac{2\pi r}{T}$ leads to an equation: $\frac{T^2}{r^3} = \frac{4\pi^2}{GM}$ which directly links a satellite's time period, T, to its distance from the centre of the mass, M, which it is orbiting. (This equation need not be remembered.)
- The speed and time period of a satellite depends only on the radius of the orbit and the magnitude of the mass being orbited, but *not* the mass of the satellite.
- Launching a satellite requires transferring to it the necessary gravitational potential energy to get it to the required height, plus the necessary kinetic energy so that it has the right speed for an orbit at that height.

Solving problems involving the speed required for an object to go into orbit around a planet and for an object to escape the gravitational field of a planet

Key concepts

Escape speed is the theoretical minimum speed required for any mass to escape the gravitational field of a planet (or moon, or star). It depends on the mass of the planet and the distance of the mass from its centre.

The *orbital speed* a satellite needs in order to maintain a circular path also depends only on its orbital radius and the mass of the planet.

Expert tip

The escape speed from the Earth's surface is 11.2 km s⁻¹. But remember that this is a theoretical value because energy and speed would be dissipated by air resistance. An object on the Earth's surface already has kinetic energy because it is moving with the rotation of the Earth's surface. This speed is about 460 m s⁻¹ on the equator, but it decreases with movement north or south because such positions have a smaller radius from the Earth's axis.

QUESTIONS TO CHECK UNDERSTANDING

26 a Confirm that the escape speed from the Earth's surface is 11.2 km s⁻¹.
 b Explain why all masses have the same escape speed.
 c Determine the escape speed from a planet which had twice the radius of Earth, but the same density.

27 The escape speed of the Moon is 2.4 km s⁻¹.
 a If the mass of the Moon is 7.3×10^{22} kg, use its escape speed to determine its radius.
 b Explain why the escape speed from the Moon is much lower than from the Earth.

28 a Calculate the orbital speed needed for a satellite to orbit the Earth at a height of 200 km.
 b What is the time period of this orbit?

29 Mars has a radius of 3390 km and a mass of 6.4×10^{23} kg. It has a day which is 40 minutes longer than Earth's. Determine the radius of the orbit of a satellite around Mars which always remains above the same place on the planet's surface.

30 The radius of the Earth is 6.4×10^6 m. Determine the minimum theoretical speed which a mass would need if, when projected vertically upwards it was to reach a height of 500 km. Assume there was no air resistance.

■ Orbital energy

- A satellite in orbit has both kinetic energy, E_K, and gravitational potential energy, E_P.
- Since $E_K = \frac{1}{2}mv_{orbit}^2$ and $v_{orbit} = \sqrt{\dfrac{GM}{r}}$, $E_K = \frac{1}{2}\dfrac{GMm}{r}$
- Reminder: $E_P = \dfrac{-GMm}{r}$.
- The total **orbital energy** of a satellite in circular orbit, $E_T = E_K + E_P = -\frac{1}{2}\dfrac{GMm}{r}$
- The potential energy and the total energy must always be negative (because infinity is chosen to have zero potential energy), but the kinetic energy is always positive. The magnitude of the kinetic energy of an orbiting satellite is equal to the magnitude of its total energy. See Figure 10.24.

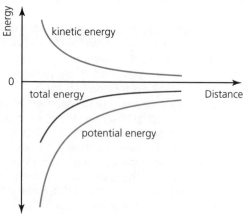

Figure 10.24

- Satellites which are in relatively low orbits around the Earth may experience some slight air resistance. *If* an unpowered satellite could stay at the same height this would result in a reduction of speed, but (as it *starts* to lose speed) the satellite moves closer to the Earth, to a smaller radius, *r*, because its total energy ($-\frac{1}{2}\dfrac{GMm}{r}$) has reduced (to a larger negative value) due to energy dissipation. The orbital speed at a lower height is *greater* than before. (This will increase air resistance further.)

■ Solving problems involving orbital energy of masses in circular orbital motion

QUESTIONS TO CHECK UNDERSTANDING

31 A 250 kg satellite is orbiting the Earth with a radius of 6.68×10^6 m.
   a Calculate: **i** its orbital speed, **ii** its kinetic energy.
   b Write down: **i** the gravitational potential energy of the satellite, **ii** the total energy of the satellite.

32 A satellite in orbit around the Earth is moved to a higher orbit. Its mass is unchanged. Compare the orbital speeds, kinetic energies, potential energies and total energies of the satellite in the two orbits.

33 Titan is the largest moon of Saturn. It has a mass of 1.3×10^{23} kg, its average distance from the centre of Saturn is 1.2×10^6 km and its orbital period is 15.9 days.
   a Calculate the orbital speed of Titan.
   b Hence determine the kinetic energy of this moon.
   c What is the gravitational potential energy of Titan?

Expert tips

There are many factors affecting the choice of height for a satellite around the Earth. All satellites need to be above the Earth's atmosphere to avoid air resistance, and lowest orbits are at a height of about 200 km. We have seen that the radius affects the time period for each orbit and the quickest (lowest) take about 90 minutes for each orbit. A satellite in a low orbit is ideal for observing the Earth's surface and it will pass over the polar regions and large parts of the world every day.

Satellite orbits can be synchronized with the Earth's rotation by placing them at the right height (about 30 000 km) so that they have a period of 24 hours. This makes them useful for communications, especially if they remain above the same place on the Earth's surface in a **geostationary orbit**. To do this they must orbit in the plane of the equator.

Radial electrostatic fields

- The gravitational field concepts discussed in the previous section can be applied by *analogy* to electrostatic fields. For the sake of clarity they are repeated below (as applied to electrostatic fields), but electrostatic fields are also more complicated because there are two kinds of charge (positive and negative).

■ Forces and inverse square law behaviour

- Consider Figure 10.25, which shows a sphere with charge $-q_1$, around which there is a radial electric field. The electric field lines are pointing inwards towards the negative charge. From Section 5.1, we know that if another small charge, $+q_2$, is at point P it will experience an electric force $F_E = \dfrac{kq_1q_2}{r^2}$ towards q_1 (and q_1 will experience an equal and opposite force). The negative sign produced in the value of F_E indicates an attractive force. If the second charge was negative, the magnitude of the forces would be the same but in the opposite directions (repulsion).

- Similar comments can be applied to the forces on charges around a *positively* charged sphere.

- A graphical representation of the inverse square variation of electric force with distance can be seen in Figure 10.26 below.

- *Electric field strength, E,* also follows an inverse square law variation with distance around a point charge, or charged sphere.

- Electric field strength, $E = \dfrac{F_E}{q}$, so that at point P, $E = \dfrac{kq_1}{r^2}$.

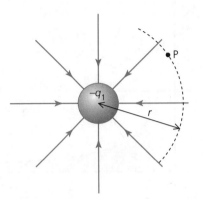

Figure 10.25

■ Solving problems involving forces on charges in radial fields

QUESTIONS TO CHECK UNDERSTANDING

34 Make a copy of Figure 10.25, but amend it so that the charge at the centre is $+q_1$. Then re-write the accompanying paragraph to describe the forces involved when another charge of $+q_2$ is placed at P.

35 What is the separation of two point charges each of 3.2 nC if the force between them is 9.2×10^{-6} N?

36 A charge of +1.2 nC is placed midway between fixed point charges of −23 nC and −34 nC which are 25 cm apart. Determine the resultant force on the 1.2 nC charge.

■ Potential energy

- Electric potential energy is stored in a system of two or more separate charges. We will often be considering situations in which a relatively small charge is moved in the field of a much larger fixed charge, and under such circumstances the smaller charge will be considered to have no effect on the larger charge, so that we may refer to the potential energy of the small charge (rather than the system).

- We have seen in Section 10.1 that the *electric potential energy, E_P,* of a charge at any point is the work done to bring that charge from infinity to that point.

- If the force is attractive (between opposite signs) then the electric potential energy will be negative because infinity is taken to be the zero of potential energy. If the force is repulsive (similar signs) the energy will be positive.

- In Section 2.3 we learned that work done can be determined from force multiplied by distance but, if the force varies, an *average* value must be used in the calculation. From Section 2.3, we also know that work done can be determined from the area under a force–distance graph.

■ Determining the potential energy of a point charge

■ Electric potential energy, E_P, may be determined from the (shaded) area under a graph like that in Figure 10.26, where r is the distance from the fixed charge to the point charge being considered.

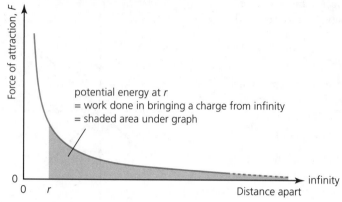

potential energy at r
= work done in bringing a charge from infinity
= shaded area under graph

Figure 10.26

■ The electric potential energy stored between two charges (points or spheres), q_1 and q_2, may be determined directly using the equation: $E_P = \dfrac{kq_1q_2}{r}$ (where r is the distance between their centres). There is an inverse relationship between gravitational potential energy and the distance from a point mass (*not* inverse square).

■ Solving problems involving potential energy

QUESTIONS TO CHECK UNDERSTANDING

37 a Calculate the electric potential energy of a hydrogen atom of radius 5.3×10^{-10} m.

 b Explain why the energy is considered to be negative.

38 A small charge of +4 nC is moved from a long way away to a point close to a fixed, positively charged sphere (100 nC).

 a Determine values for the forces on the charges and thereby draw a force–distance graph which will enable you to determine the change in electric potential energy when the charge moved from 20 cm to 10 cm from the centre of the sphere.

 b Is this energy positive or negative? Explain your answer.

 c Use the equation to calculate the potential energy of the small charge when it is 10 cm from the sphere.

■ Potential

■ We know that the *electric potential*, V_e, at a point is defined as the work per unit charge that would need to be done to bring a small test positive charge from infinity to that point. It can be calculated for a charge at any point by dividing the electric potential energy by the charge, q: $V_e = \dfrac{E_P}{q} = \dfrac{kq}{r}$.

■ Figure 10.27 shows, on the same axes, the variation of potential with distance around a positively charged sphere (or point) and also around a negatively charged sphere (or point) which has charge of the same magnitude. There is an inverse relationship between potential and distance (*not* inverse square).

■ Electric equipotentials around points (or charges on spheres) were shown in Section 10.1, Figure 10.10.

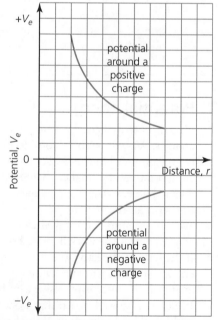

Figure 10.27

Expert tip

Potential wells (see Figure 10.19) provide a useful visualization for electric fields as well as gravitational fields. For example, we may consider that an electron in a hydrogen atom moves around in the potential well produced by the proton in the nucleus.

■ Combining fields and potentials

- A charge may be in two electric fields, such as provided by the two charged spheres shown in Figure 10.28. A positive charge placed at point P will experience forces from both spheres, as shown.
- Electric force and field strength are vector quantities and the combined force or field may be determined by vector addition.
- Electric potential energy and potential are scalars, and the combination of two energies or potentials can be determined by simple addition.

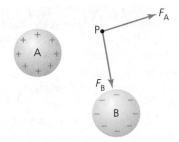

Figure 10.28

■ Determining the potential inside a charged sphere

- If a *hollow spherical conductor* is charged, the charges will move to the outer surface because their mutual repulsion pushed them as far apart as possible. See Figure 10.29 which shows a negatively charged sphere.
- There is zero electric field inside a charged sphere, which means that the potential gradient is also zero, so that the potential is constant, and equal to that on the conductor. See Figure 10.30.
- Similar comments apply to other shaped conductors with enclosed spaces (like Faraday cages).

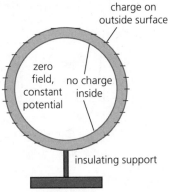

Figure 10.29

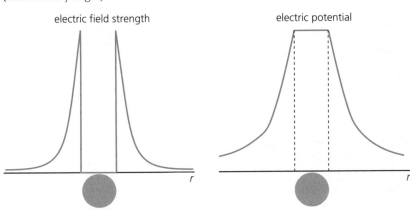

Figure 10.30

■ Summary of the four inter-connected radial electric field concepts

Force	Potential energy
$F_E = k\dfrac{q_1 q_2}{r^2}$	$E_p = \dfrac{kq_1 q_2}{r}$
Field	Potential
$E = \dfrac{kq}{r^2}$	$V_e = \dfrac{kq}{r}$

Figure 10.31

■ Potential difference

- This has already been discussed in Section 10.1. Reminder: the energy (work), W, required to move a charge, q, between points which have a potential difference of ΔV_e can be determined from $W = q\Delta V_e$.

■ Potential gradient

- The gradients of potential–distance graphs, like those shown in Figure 10.27, are $\dfrac{\Delta V_e}{\Delta r}$. These are called electric *potential gradients* and they are important because the electric potential gradient at any point is equal to the electric field strength, E at that point: $E = \dfrac{-\Delta V_e}{\Delta r}$

> **Key concept**
>
> **Electric potential gradient** is equal to the electric field strength, E at that point: $E = \dfrac{-\Delta V_e}{\Delta r}$.

QUESTIONS TO CHECK UNDERSTANDING

39 Calculate the electric potential at a distance of 10.0 cm from the surface of a sphere of radius 5.0 cm if it has a charge of −4.5 pC.

40 Consider Figure 10.28. If the charge on A is 12 nC and the charge on B is −2.4 nC, determine the potential at point P, which is 15 cm from the centre of A and 10 cm from the centre of B.

41 Figure 10.32 shows the variation of potential with distance from a point charge.

   a Is the charge positive or negative?

   b Use the graph to estimate the strength of the electric field at a distance of 15 cm from the charge.

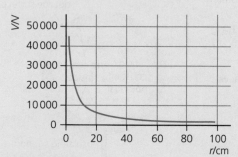

Figure 10.32

42 Consider Figure 10.12.

   a What is the potential difference between the +40 V and −40 V equipotentials?

   b Estimate how much energy would be transferred when moving a 2 nC charge from the surface of one sphere to the surface of the other.

   c Discuss whether the energy is positive or negative.

Uniform electrostatic fields

- The uniform electric field between parallel charged metal plates was introduced in Section 10.1. We will now look at this situation in more detail. Consider Figure 10.33, which shows a charge q in the **uniform electric field** created by a potential difference of ΔV_e across plates separated by a distance r.
- If the charge moves from one plate to the other, the work done, $W = Fr$ (force × distance), and from the definition of p.d., we also know that $W = q\Delta V_e$.
- Therefore, $\dfrac{F}{q} = \dfrac{\Delta V_e}{r}$.
- Electric field strength, E, is defined as $\frac{F}{q}$, so that we now see that $E = \dfrac{\Delta V_e}{r}$ (unit: $V\,m^{-1}$) is an alternative, and much more convenient, way of measuring a uniform electric field.
- This is just a particular example of the more generalised interpretation of field strength as potential gradient that we have met before $\left(E = \dfrac{-\Delta V_e}{\Delta r} \right)$.

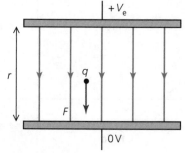

Figure 10.33

Motion of charged particle beams in uniform fields

- Forces on atomic particles in gravitational fields are so small that they can be ignored. But electric and magnetic forces can be strong enough to affect the motion of charged particles. The deflection of particle beams in fields of known strength can be used to determine properties of the particles. The beams of charged particles that could be discussed in this course are electrons (including beta particles), protons and alpha particles.
- Figure 10.34 shows the design of an electron deflection tube which can used to investigate the deflection of an electron beam in vertical and horizontal

uniform *electric fields*. Electrons are emitted from the hot cathode and accelerated by the positive potential on the anode. They move at high speed across the evacuated tube and hit the fluorescent screen at the end of the tube.

■ Kinetic energy gained by electron, $E_K = \frac{1}{2}mv^2$ = charge × p.d. = eV

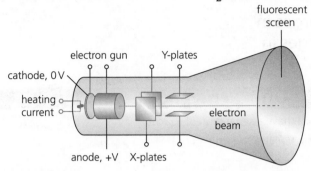

Figure 10.34

■ A p.d. applied across the X-plates and/or the Y-plates can deflect the electron beam. An electron initially moving parallel to the plates will experience a force perpendicular to the plates, as described in the previous sub-section. The electron will accelerate towards the positive plate, while at the same time continuing with a constant speed along the tube. This results in a parabolic trajectory.

■ The motion of charged particles across uniform *magnetic fields* has been briefly described in Section 5.4. If charges move perpendicularly to the field, they experience a force which is always perpendicular to their motion, and this provides the centripetal force needed for them to move in the arc of a circle.

■ Equating the expression for centripetal force (Section 6.1) to the force on a charge moving perpendicular to a magnetic field (Section 5.4) we get $\frac{mv^2}{r} = qvB$, which leads to an expression for the radius, r of the circular path of a charge Q moving across a magnetic field of strength B: $r = \frac{mv^2}{qB}$. This is a useful equation, but it need not be remembered.

■ Figure 10.35 shows the circular paths of positive and negative charges of different speeds passing across magnetic fields.

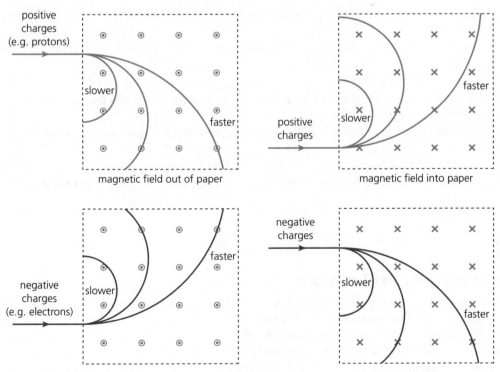

Figure 10.35

> **Expert tip**
>
> Strong magnetic fields provide the only means for making charged particles move in circular paths. Applications of this theory include *mass spectrometers* for the identification of nuclides and *particle accelerators*.

■ Solving problems involving forces on charges in uniform fields and orbital energy of charges in circular orbital motion

QUESTIONS TO CHECK UNDERSTANDING

43 A p.d. of 1000 V is connected across parallel metal plates which have a separation of 3.2 cm.

   a   What is the electric field strength between the centres of the plates?

   b   Estimate the field strength at the edge of the plates.

   c   What force would be experienced by a +1.0 nC charge midway between the centres of the plates?

   d   What affect does this force have on the charge?

   e   How does the force change if the charge moves closer to one of the plates?

44 A 20 cm long, straight cylindrically shaped metallic conductor has a p.d. of 12.8 mV across its ends.

   a   What is the value of the electric field strength along the conductor?

   b   Determine the magnitude of the force acting on free electrons in the metal.

   c   Determine the magnitude of the acceleration produced.

45 An electron beam was directed horizontally and perpendicularly across the uniform electric field between parallel plates. The plates had a p.d. of 4000 V across them and they were separated by 3.4 cm.

   a   Draw a sketch of this arrangement.

   b   The beam was produced when electrons were accelerated from rest by a p.d. of 5000 V. Determine the kinetic energy gained by an electron in: **i** electronvolts, **ii** joules.

   c   Calculate the maximum speed of the electrons in the beam.

   d   If the plates were 6.0 cm long, what was the time taken for an electron to pass between them?

   e   Calculate the magnitude of the electric field between the plates.

   f   Determine the force on each electron.

   g   Calculate the acceleration of the electrons.

   h   Determine the deflection of the beam as it passed between the plates.

46 a   An alpha particle moving with a speed of 6.1×10^6 m s^{-1} enters a perpendicular magnetic field of strength 0.46 T in a vacuum. Determine the radius of its path.

   b   What would be the radius of the path of an alpha particle travelling at one tenth of the speed which passed through the same field?

   c   What is the energy of the alpha particle (in MeV)?

   d   How would the path of the alpha particle be affected if air at low pressure was introduced into the experiment? (Mass of alpha particle = 6.64×10^{-27} kg)

47 a   Sketch an arc of the circular path followed by a negative charge, $-q$, and mass m, moving perpendicularly across a magnetic field strength B.

   b   Add a sketch of the path followed by the same charge passing across a field of strength $2B$.

   c   Add a third sketch showing the path followed by the path of a charge $-q$ and mass $2m$ passing through a field of strength B.

NATURE OF SCIENCE

■ Communications of scientific explanations

The *field* concept is used by scientists to help develop models to explain phenomena as wide-ranging as the motion of distant galaxies and the behaviour of subatomic particles. Many of the general public have a natural curiosity about scientific theories on subjects such as these, but understanding them can be very difficult for non-scientists in an increasingly complex world. Scientists, and the organizations they work for, have a responsibility to inform the public, but this requires considerable communication skills in a world that usually requires information in short and easily understood news items.

11 Electromagnetic induction

11.1 Electromagnetic induction

Revised ☐

Essential idea: the majority of electricity generated throughout the world is generated by machines that were designed to operate using the principles of electromagnetic induction.

Electromotive force (emf)

Revised ☐

■ The concept of *emf* was introduced in Chapter 5. We usually use this term to describe the electric potential difference produced by a process or device which can convert some other form of energy into electrical energy. (More precisely, emf is the total energy transferred per coulomb by the process.)

Describing the production of an induced emf by a changing magnetic flux and within a uniform magnetic field

Revised ☐

> **Key concepts**
>
> **Electromagnetic induction** is the process by which an emf is induced (made to occur) when a conductor experiences a changing magnetic field.
>
> Electromagnetic induction may involve a moving magnet, a moving conductor or the magnetic field around a changing current.

■ (The term *magnetic flux* is explained later.)
■ Electromagnetic induction can be demonstrated by moving a permanent magnet near a stationary conductor, or, conversely, by moving a conductor through a stationary magnetic field (see Figure 11.1). If the conductor and the magnetic field are *both* moving with the same velocity, no emf will be induced: *relative* motion between them is needed.

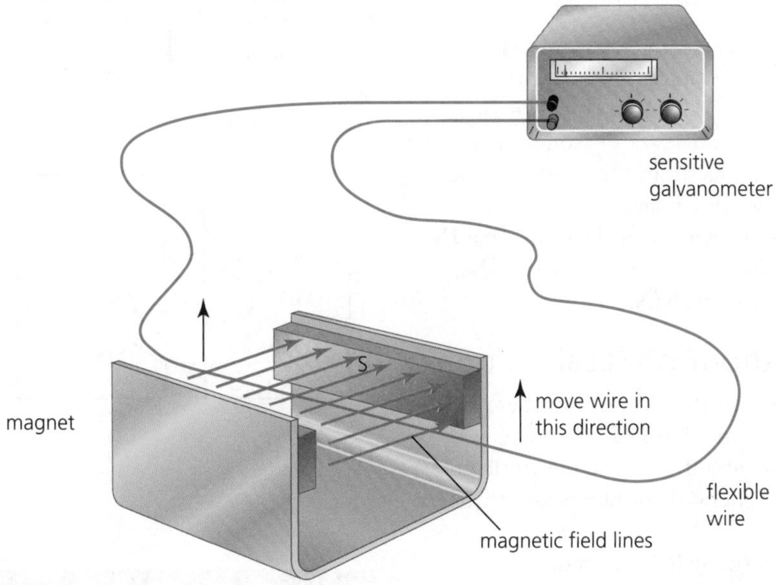

Figure 11.1

■ If a conductor is in a complete circuit, an *induced emf* will result in an *induced current*.
■ The greatest emf is induced when the motion of the conductor is *perpendicular* to the direction of the magnetic field (or vice versa). No emf is induced if the conductor is moved parallel to the field.
■ The magnitude of the induced emf increases with the speed of relative motion, the strength of the magnetic field and the number of *turns* in the circuit.

> **Expert tip**
>
> If a number, N, of separate wires were moved through the field in Figure 11.1, the same emf, ε, would be induced across each of them. If instead they are connected end-to-end with the connecting leads *outside* of the field, then the emfs will add up to induce a total emf of $N\varepsilon$. We would describe this arrangement as a *coil* of N turns.

- Using coils with large numbers of turns is the usual way of increasing the magnitude of induced emfs. Figure 11.2 shows a demonstration of electromagnetic induction as a permanent magnet is moved into, or out of, a coil. The direction of the induced emf and current depend on which way the magnet is moving. Reversing the polarity of the magnet will reverse the direction of the induced emf and current.

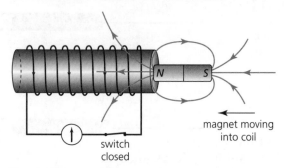

Figure 11.2

- The rotation of coils in magnetic fields is the means by which most of the world's electrical energy is generated. See Section 11.2.
- To understand electromagnetic induction, consider the simplest possible arrangement: Figure 11.3 shows a conductor moving at a speed v (towards the bottom of the page), perpendicularly across a uniform magnetic field of strength B. The electrons within the conductor are charges moving across a magnetic field and will therefore experience magnetic forces, as explained in Section 5.4. The direction of this force can be determined using the *left-hand rule*, remembering that the direction of conventional current is opposite to the direction of electron movement.
- The force on one electron is shown in the figure. The movement of electrons results in one side of the conductor gaining negative charge, and the other side losing negative charge and so becoming positive. In this way an emf is produced across the conductor.
- The movement of charge does not continue indefinitely because further electrons moving towards the negatively charged side of the conductor are repelled (and attracted to the other end).
- We can use this fact to determine an equation for the induced emf. At equilibrium, the electric repulsive force = magnetic force: using equations from Chapter 5, $Eq = qvB$ (in this arrangement $\sin \theta = 1$). This leads to $E = Bv$.
- We know from Chapter 10 that a uniform electric field, $E = \frac{\Delta V}{\Delta r}$ or, in this arrangement, $E = \frac{\varepsilon}{l}$, where l is the length of conductor in the magnetic field and ε is the induced emf. Hence the equation for the induced emf when a straight conductor moves *perpendicularly* across a magnetic field becomes $\varepsilon = Bvl$.
- If the circuit has N turns, the same emf is induced across each turn, so that the equation for the total emf induced becomes $\varepsilon = BvlN$.

▣ Electromagnetic induction without physical motion

- The necessary requirement for electromagnetic induction is that there is a changing magnetic field passing through a conductor. This can occur *without* physical motion if the current in one circuit (and the magnetic field around it) is changing, and the changing magnetic field then passes through another circuit.
- Figure 11.4 shows two separate circuits. When the switch in circuit A is open there are no currents and no magnetic fields. *But at the moment* when the current in circuit A is switched on, the magnetic field around it passes though circuit B, inducing an emf and a current through the meter. There is no induction when the current and field are constant. The induced emf and current are induced in the opposite direction *at the moment* that the switch is turned off. The induced emf can be made much larger by using coils of wire.
- If the battery is replaced with an alternating current supply, electromagnetic induction will be continuous, producing an alternating induced emf. This effect is used in transformers; see Section 11.2.

> **Key concept**
>
> Electromagnetic induction occurs because free electrons within the conductor experience magnetic forces (see Section 5.4). This results in some charge separation in the conductor.

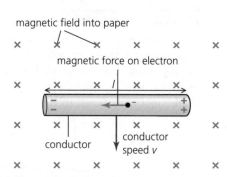

Figure 11.3

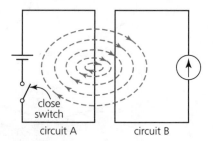

Figure 11.4

> **Expert tip**
>
> In principle, *everywhere* that alternating currents flow, oscillating electromagnetic fields will radiate away from the circuits, with the ability to induce emfs in any conductors through which they pass. Wireless communication is just one application of this phenomenon.

QUESTIONS TO CHECK UNDERSTANDING

1 Describe how the wire shown in Figure 11.1 could be moved without inducing an emf.

2 Why are conductors (rather than insulators) needed in order to produce electromagnetic induction?

3 Consider Figure 11.2. List three ways in which the induced emf could be increased in this kind of experiment.

4 Figure 11.5 shows a permanent magnet being dropped through a coil of wire connected so that the emf induced can be recorded electronically and a graph drawn.

   a Explain why a data logger is needed for this experiment.

   b Sketch a graph to show how the induced emf changes as the magnet passes through the coil.

5 A conductor of length 50 cm moved horizontally across a vertical magnetic field of strength 45 mT with a speed of 32 cm s⁻¹. What emf was induced across the conductor?

6 Figure 11.6 shows an arrangement of two separate circuits containing coils of wire with the same iron core. The apparatus is to be used to demonstrate electromagnetic induction.

   a Explain why an iron core is being used.

   b Explain how turning on a *direct* current in circuit A results in a momentary deflection on the meter connected to circuit B.

   c Describe the emf induced in circuit B if an *alternating* voltage is applied to circuit A.

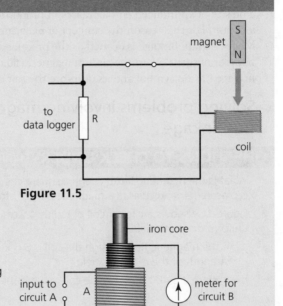

Figure 11.5

Figure 11.6

Magnetic flux and magnetic flux linkage

■ The very wide variety of circumstances in which emfs can be induced can be confusing, but they are better understood by first understanding the concept of *magnetic flux, Φ*.

■ If a magnetic field is *perpendicular* to an area, *magnetic flux* is determined by multiplying the magnetic field strength by the area, $\Phi = BA$.

■ Figure 11.7 shows a more general situation, in which the flux is at an angle θ to a normal to the surface

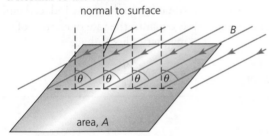

Figure 11.7

■ In unscientific terms, magnetic flux may be considered as a measure of the total magnetic field passing through an area. An understanding of magnetic flux may be gained by counting magnetic field lines. Consider Figure 11.8.

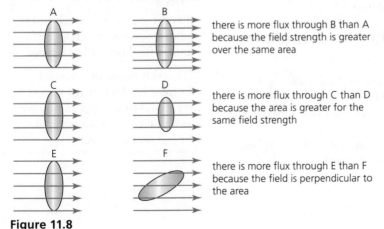

there is more flux through B than A because the field strength is greater over the same area

there is more flux through C than D because the area is greater for the same field strength

there is more flux through E than F because the field is perpendicular to the area

Figure 11.8

> ### Key concepts
>
> The **magnetic flux**, Φ, through an area is defined as the product of the component of the magnetic field strength perpendicular to that area ($B\cos\theta$) and the area A.
>
> $$\Phi = BA\cos\theta$$
>
> **magnetic flux linkage** = $N\Phi$

> ### Expert tip
>
> An analogy with radiation falling on a solar panel may also be helpful: combining intensity of radiation (solar flux density), angle of incidence and the area of the panel can lead to the determination of the total energy (flux) incident on the panel.

- The units of magnetic flux are **webers**, Wb. $1\,\text{Wb} = 1\,\text{T} \times 1\,\text{m}^2$.
- The size of an induced emf increases in proportion to the number of turns in a circuit. For this reason the concept of *magnetic flux linkage* is introduced. Magnetic flux linkage is defined as the product of the magnetic flux and the number of turns, N, in the circuit: magnetic flux linkage = $N\Phi$. Magnetic flux linkage has no symbol and its units are the same as for magnetic flux (Wb).

> **Key concept**
>
> Magnetic field strength, B, can be interpreted as magnetic flux per unit area and, as such, can be described as **magnetic flux density**

Solving problems involving magnetic flux, magnetic flux linkage

QUESTIONS TO CHECK UNDERSTANDING

7 What is the magnetic flux through a square coil with sides of 5.0 cm when it is placed perpendicular to a magnetic field of strength 35 mT?

8 Figure 11.9 shows a circular coil of radius 4.2 cm (seen from the side) in a uniform magnetic field.

   a If the magnetic flux through the coil is $5.6 \times 10^{-6}\,\text{Wb}$, what is the strength of the magnetic field?

   b If there are 480 turns on the coil, what is the flux linkage through it?

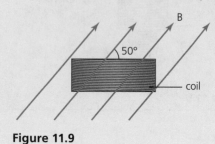

Figure 11.9

Faraday's law of induction

Revised ▢

- Faraday realized that *all* examples of electromagnetic induction can be described by one 'simple' equation.
- The negative sign in this equation is explained below (see Lenz's law).
- Remembering that $\Phi = BA\cos\theta$, we can re-write the equation in two alternative ways:
  - $\varepsilon = -NA\cos\theta \times \left(\frac{\Delta B}{\Delta t}\right)$ for circumstances when there is a changing magnetic field passing through a fixed arrangement (constant A)
  - $\varepsilon = -NB\cos\theta \times \left(\frac{\Delta A}{\Delta t}\right)$ for circumstances when there is movement of a conductor across a fixed magnetic field. For example, if a conductor of length l moves at speed v perpendicularly across a field of strength B, then $\cos\theta = 1$ and $\left(\frac{\Delta A}{\Delta t}\right) = vl$, so that the equation reduces to $\varepsilon = -BvlN$, which is the same as we met earlier in this section (but now with a negative sign included).

> **Key Concept**
>
> **Faraday's law of induction:** the magnitude of any induced emf is equal to the rate of change of magnetic flux linkage.
>
> $$\varepsilon = -N\frac{\Delta\Phi}{\Delta t}$$

Lenz's law

- Lenz's Law is represented mathematically by the inclusion of the negative sign in the equation for Faraday's law of induction (above).

Explaining Lenz's law through the conservation of energy

- Any induced current must have a magnetic field around it (as discussed in Section 5.4). The direction of that magnetic field is very important. As we shall explain, the direction of the induced field is *opposite* to the original field involved in the electromagnetic induction.
- Any induced current will involve energy and we need to be able to explain where that energy has come from. It must have been transferred to the induced current from somewhere else.

> **Key Concept**
>
> **Lenz's law:** An induced emf always opposes the change of magnetic flux which produced it.
>
> This is because any induced current has had energy transferred to it, and this energy must have been taken from the original energy in the system.

- The simplest induction example is of a conductor moving into or out of a magnetic field (see Figure 11.1). The energy of the induced current is transferred to it from the kinetic energy of the conductor, which will experience a retarding force that makes it slow down. The energy lost by the conductor is transferred to the induced current. If the conductor is kept moving at a constant speed, then the energy will be transferred from the work done against the retarding force.
- The conductor experiences a retarding force because the current is in such a direction that the magnetic field around it is in opposition to the original field provided by the magnet. In other words, the induced current direction is *always* such that it sets up a magnetic field opposing motion.
- Figure 11.2 provides another example: when the magnet is pushed into the coil it induces a current which makes the coil behave as a magnet (see Section 5.4). In this case, the current flows around the coil in a direction such that the end of the coil closest to the permanent magnet is a *north* pole. The magnet is therefore repelled and slowed down (unless it is pushed). The magnet loses (kinetic) energy, which is transferred to the current. If the magnet is moved away from the coil, the induced current will oppose its motion by flowing in the opposite direction, which makes the end of the coil a *south* pole.

■ Solving problems involving Faraday's law

- In this course, the equation $\varepsilon = -\frac{N\Delta\Phi}{\Delta t}$ may be applied to (1) straight conductors moving at right angles to magnetic fields, (2) rectangular coils moving in and out of uniform magnetic fields, (3) rectangular coils rotating in uniform magnetic fields.

QUESTIONS TO CHECK UNDERSTANDING

9 A train is moving in a straight line at a speed of 2.0 m s⁻¹ across the Earth's magnetic field, which has a strength of 35 μT at that location. Determine the emf induced across an axle of the train if its length is 1.42 m and the magnetic field is at an angle of 70° to the vertical.

magnetic field into paper

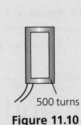

× × × × × ×
× × × × × ×
× × × × × ×   P
× × × × × ×   •
× × × × × ×

10 Figure 11.10 shows a rectangular coil of size 4.3 cm × 2.7 cm and 500 turns placed outside of a uniform magnetic field of strength 5.6 mT.

500 turns

Figure 11.10

a If the coil is moved to the centre of the field in 0.50 s, what is the magnetic flux through the coil in that position?

b Determine the magnitude of the average emf induced during the movement.

c Determine the magnitude of the average emf produced if the coil is then turned over in a time of 0.50 s.

d Sketch a graph to show how the induced emf changes as the coil is moved from its original position at a constant speed to point P (no values needed).

11 Figure 11.11 shows a rectangular loop of wire which can be made to rotate at a constant speed within a uniform magnetic field. The dimensions of the coil are 6.2 cm × 3.6 cm.

a Explain why no emf will ever be induced across XY.

b In which position of the loop will the maximum emf be induced?

c What is the speed of the sides of the loop when it rotates with a frequency of 10 Hz?

d Calculate the maximum emf produced across the loop at this frequency if the field strength is 0.26 T.

e Determine the emf being induced when side XY is at: **i** 20°, **ii** 70° and **iii** 90° to the horizontal.

f How many turns would be needed for a coil of these dimensions to induce a maximum emf of 1.0 V at a frequency of 50 Hz?

g Explain why this apparatus can be described as an *ac* generator.

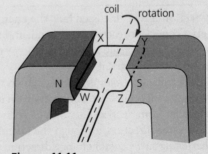

Figure 11.11

12 Figure 11.12 shows a rod AB moving to the right across a perpendicular magnetic field, *B*, with a speed *v*. The rod can move without friction along the parallel conductors, which complete an electrical circuit as shown.

  a Electrons in the rod experience magnetic forces as they pass through the field. What is the direction of these forces?

  b In which direction does conventional current, *I*, flow around the circuit?

  c Explain why a force is needed to keep the rod moving at a constant speed.

  d Write down an expression (from Section 5.4) for that force.

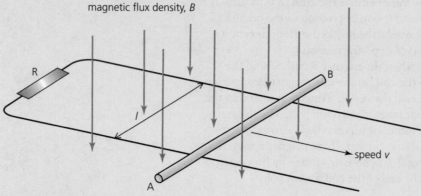

Figure 11.12

13 Two magnets are dropped from the same height to the ground. One magnet falls freely, the other falls through a coil connected in a circuit, as shown in Figure 11.5.

  a Explain why the magnets take different times to reach the ground.

  b Explain why the times would be the same if the switch was opened.

NATURE OF SCIENCE:

■ Experimentation

Michael Faraday is rightly seen as one of the greatest of scientists. His pioneering work on electromagnetism was based on excellent experimental skills: making observations (with basic apparatus) that most others would regard as insignificant, or even experimental errors.

11.2 Power generation and transmission

Revised ☐

Essential idea: Generation and transmission of alternating current (ac) has transformed the world.

Alternating current (ac) generators

Revised ☐

- In Section 8.1 we discussed how the energy transferred from (1) burning fuels, (2) nuclear reactions, (3) wind, and (4) falling water could be used to turn *turbines* to create the kinetic energy that is necessary for the large-scale generation of electrical energy.
- An electrical *generator* is a device which transfers kinetic energy to electric energy. In this course we concentrate on *alternating (ac) current generators*.
- Reminder from Section 5.1: *direct currents* (dc) always flow around a circuit in the same direction, although the magnitude of the current may vary. *Alternating currents* (ac) change direction periodically.

■ Explaining the operation of a basic ac generator, including changing the generator frequency

- We have seen in Section 11.1 that the side of a coil moving across a magnetic field will have an emf induced across it. If both sides of a coil move in the same

> **Key concepts**
>
> Alternating currents are generated in coils rotating in magnetic fields. In an **ac generator**, brushes and slip rings enable the current to flow into an external circuit.
>
> If a certain coil rotates at a higher frequency, the induced emf will be greater.

direction through the same field, the two emfs will cancel because they are equal and opposite, but if the coil *rotates* within a magnetic field, the emfs induced on opposite sides will combine, because as one side goes up the other side goes down. Figure 11.13 shows a simple ac generator (but, for clarity, it only shows one loop of a coil).

■ If the coil rotates at a constant speed, the induced emf and current will have *sinusoidal* waveforms as shown in Figure 11.14.

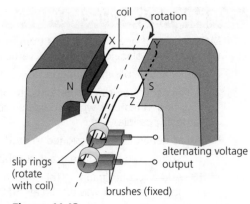

Figure 11.13

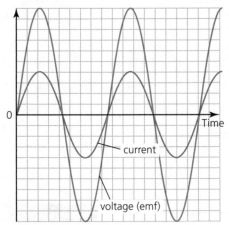

Figure 11.14

■ The shape of the graphs in Figure 11.14 may be understood by considering the instantaneous motion of the coil at different positions during each rotation. See Figure 11.15.

■ When the coil is horizontal (position **a**) the sides WX and YZ are moving through the horizontal field at the quickest rate, so that the induced emf has its greatest value. In position **b** the sides are instantaneously moving parallel to the field, so that the induced emf is zero. Position **c** is similar to **a**, but the sides are now moving in the opposite direction (compared to **a**), so that the emf is reversed.

■ If the speed of rotation increases, the induced emf will have a greater frequency *and* a greater amplitude (because the rate of change of magnetic flux is greater. See Figure 11.16.

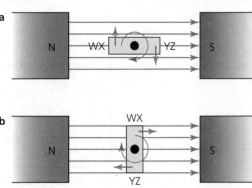

Figure 11.15

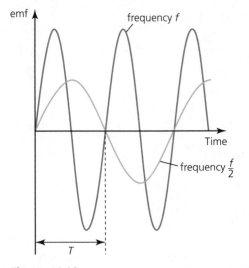

Figure 11.16

■ Practical generators may have more than one coil and the coil(s) will be wound on iron cores. They may also use electromagnets instead of permanent magnets.

QUESTIONS TO CHECK UNDERSTANDING

14 Make a copy of one of the graphs from Figure 11.16 and mark on it every time when the plane of the coil was vertical or horizontal.

15 Explain why the coils in generators are usually wound on iron cores.

16 a   An ac generator is being designed to operate at a fixed frequency. List four design features which could be used to make the output emf as large as possible.

   b   An ac generator was designed to produce 110 V 50 Hz.

      i   Sketch a voltage–time graph for this generator over two cycles.

      ii Add a second graph on the same axes to show the output if the same generator was turned at 25 Hz.

17 A small ac generator has a 600 turn coil with an area of 5.6 cm². What is the magnitude of the average emf produced if it is rotating with a frequency of 10 Hz in a uniform magnetic field of strength 0.40 T?

Average power and root mean square (rms) values of current and voltage

Revised ▢

■ Figure 11.17 shows how the voltage and the current vary with time in the simple ac resistor circuit shown. They are in phase with each other and both periodically change direction, such that their true average values are zero. (See below for comments about *effective* average values.) Note the symbol for the ac power supply. The current, *I*, is always proportional to the voltage, *V* across the ohmic resistor ($V = IR$).

■ The maximum values are also called the **peak values**.

■ From Chapter 5 we know that, in general, electrical power, $P = IV$.

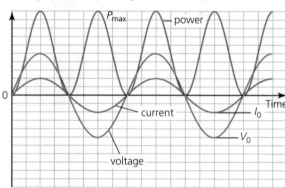

Figure 11.17

■ So, for example, the maximum (peak) power can be determined from the maximum values of current and voltage (I_0 and V_0): $P_{max} = I_0 V_0$.

■ The curve representing power, *P*, in Figure 11.17 has been determined by multiplying the instantaneous values for voltage and current. The power is always positive because *I* and *V* always have the same signs as each other. Positive power means that energy is always being transferred *to* (rather than from) the resistor, where it is transferred to internal energy.

■ For a power graph of this shape (a sine squared graph) the *average* value, $\overline{P}$, is half the peak value ($\overline{P} = \frac{1}{2}P_{max}$), so that $\overline{P} = \frac{1}{2}I_0 V_0$.

■ The *effective average values* of current and voltage are called **root mean squared (rms) values**, but there is no requirement to understand the origin of this term in the IB Physics course.

■ So that $\overline{P} = I_{rms} V_{rms}$. This means that the *rms values* for the current and voltage must be equal to their maximum values divided by $\sqrt{2}$: $I_{rms} = \frac{I_0}{\sqrt{2}}$ and $V_{rms} = \frac{V_0}{\sqrt{2}}$.

■ An electrical supply rated at, for example, 230 V has exactly the same heating effect in a resistor whether it is alternating or constant.

> **Key concept**
>
> An *rms value* is that value of an ac current (or voltage) which dissipates power in a resistor at the same rate as a steady direct current (or voltage) of the same value.

■ Ohm's law can be used with rms values. $R = \dfrac{V_{rms}}{I_{rms}} = \dfrac{V_0}{I_0}$.

■ Remember also (from Chapter 5) that $P = \dfrac{V^2}{R}$ and $P = I^2R$. These equations can be used for alternating currents with peak or rms values.

■ Solving problems involving the average power in an ac circuit

QUESTIONS TO CHECK UNDERSTANDING

18 An electrical supply is rated at 110 V ac 50 Hz.

  a What is the maximum value of this voltage?

  b What rms current will be produced when the supply is connected across a heating element of 15.0 Ω in an electric toaster?

  c What is the (average) power of the toaster?

19 Consider again Figure 11.17.

  a If the average power transformed in the resistor was 24 W, what was the maximum rate of energy transfer?

  b If the peak voltage was 12.0 V, what was the peak value of the current?

  c Determine the rms current in the circuit.

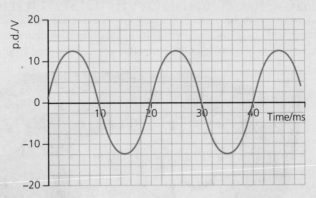

Figure 11.18

20 The resistance of a domestic water heater is 28 Ω and it is connected to an ac voltage supply, V_{rms} = 230 V.

  a Calculate the average rate at which energy is transferred to the water.

  b What dc voltage would provide the same power when connected to the same heater?

21 Figure 11.18 shows how a voltage varies with time.

  a What is its peak value?

  b What is its average value?

  c Calculate V_{rms}.

  d What is the frequency of this supply?

Transformers

Revised

■ A **transformer** is a device which changes (transforms) an alternating voltage to another value. It uses the electromagnetic induction created by a time-changing magnetic flux (see Section 11.1).

■ Figure 11.19a shows the essential features of a simple transformer. Two coils of wire are wound on the same soft iron core. Figure 11.19b shows the circuit symbol for a transformer.

> **Key concept**
>
> In a transformer an alternating voltage, ε_p, drives an alternating current, I_p, through the **primary coil** creating an *alternating magnetic flux linkage* which passes through the iron core to the **secondary coil**, where it induces an emf, ε_s.

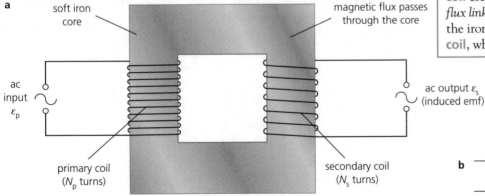

Figure 11.19

- For an *ideal transformer* in which there is no dissipation of energy to internal energy, the input power and output powers will be equal: $I_p \varepsilon_p = I_s \varepsilon_s$.
- Or, $\dfrac{\varepsilon_p}{\varepsilon_s} = \dfrac{I_s}{I_p}$. The ratio of the input and output voltages in an ideal transformer are decided by the ratio of number of turns, so that:

$$\frac{\varepsilon_p}{\varepsilon_s} = \frac{N_p}{N_s} = \frac{I_s}{I_p}.$$

Describing the use of transformers in ac electrical power distribution

Revised ☐

- Electrical energy is transferred around the world using *transmission lines*, which are thick cables made of good metallic conductors. See Figure 11.20.

Figure 11.20

- The cables need to have low resistance (from Section 5.2: $R = \dfrac{\rho L}{A}$), so they are made of metals of relatively low resistivity. Aluminium is usually used because of its low cost and density. Thinner cables of copper (for example) could equally well be used, but the cost would be higher.
- In order to minimize power losses by keeping the currents in transmission lines low, high voltages must be used if large amounts of power are to be transferred ($P = VI$). Therefore, transformers are needed to *step-up* (increase) and then *step-down* (decrease) the voltage at various places along the transmission line. See Figure 11.21.

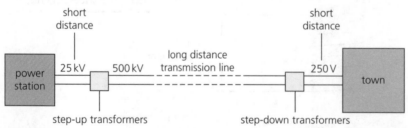

Figure 11.21

- Safety considerations will limit the maximum voltages that can be used.
- Real transformers are not 100% efficient because of power losses due to:
  - ☐ some energy is transferred to internal energy as currents flow through the resistance of the coils (sometimes called **joule heating**)

□ some energy is transferred to internal energy as **eddy currents** (see below) flow through the core, to limit this effect the core is usually *laminated* (made from many insulated layers)

□ some energy is involved when the core is repeatedly magnetized and demagnetized (**magnetic hysteresis effects**)

□ some magnetic flux also 'leaks' from the iron core.

■ Because transformers are needed, electrical power is usually transferred over large distances using *alternating currents* (dc would not produce the necessary changing magnetic flux for the operation of a transformer).

> **Key concept**
>
> In calculations we will assume that transformers are 100% efficient, but in practice some energy will be transferred to the environment, mainly due to heating of the coils and core.

Expert tip

When any piece of metal experiences a changing magnetic flux, currents will be induced and circulate *within the solid metal* (transferring energy to internal energy). These are known as *eddy currents*. The eddy currents will produce their own magnetic field which will oppose the change that produced them (an example of *Lenz's law*). Figure 11.22 shows a magnetic field perpendicular to part of a rotating metal disc. The induced eddy currents will circulate in directions such as to oppose the motion of the disc, which will slow down. The disc loses the energy that was transferred to the currents.

Figure 11.22

■ Solving problems involving step-up and step-down transformers

■ Calculations will be restricted to ideal transformers.

QUESTIONS TO CHECK UNDERSTANDING

22 A transformer designed to use an input of 230 V is used to supply an output of 12 V.

  a Is this a step-up or a step-down transformer?

  b If the primary coil has 2500 turns, how many turns are on the secondary coil?

  c If a 64 W device is connected to the output, what current will flow through it?

  d What current flows through the primary coil when:

    i   the 64 W device is connected across the secondary coil

    ii there is no device connected to the secondary?

23 Figure 11.21 shows the position of a step-up transformer in an electrical power transmission system.

  a Explain why it is located close to the power station.

  b Write down a possible turns ratio for this transformer.

  c At a time when the input current was 28 A and the transformer was 95% efficient, calculate the power loss in the transformer.

  d Write down two possible reasons for this power loss.

24 a Draw a circuit diagram in which an alternating voltage supply is connected to a step-down transformer which has a 2.5 V lamp connected across its output.

  b If the transformer has two coils with turns of 60 and 2400, what input voltage is required for the lamp to be operating correctly?

Half-wave and full-wave rectification

Revised ☐

■ Although alternating currents are used for transferring large amounts of electrical energy around the country, they are not suitable for use with many electronic devices (which need dc).

■ Therefore we need to convert ac to dc before a power source is connected to many devices. Converting ac to dc is known as **rectification**. Transformers which supply low voltages often also include a rectification circuit.

■ The simplest device which can be used to rectify a current is a *diode*. A **diode** allows current to pass through it only in one direction. Figure 11.23 shows the output from a transformer connected through a diode to a *load* (resistor). The arrow of the circuit symbol shows the direction in which conventional current can pass through the diode. We will assume that the

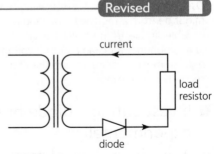

Figure 11.23

diode has insignificant resistance for current flowing in this direction, but a very large resistance for the opposite direction.

■ Figure 11.24 shows the variations in voltage across the resistor for (i) a circuit without a diode and (ii) a circuit with a diode (connected both ways). The use of a single diode produces **half-wave rectification**. *Full-wave rectification* is discussed below.

■ The shapes of the graphs in Figure 11.24 could also represent the variation of current in the circuit (because the current through an ohmic resistor is proportional to the voltage across it).

■ Ammeters and voltmeters would be of little use in this circuit because they are calibrated for use with either dc or full sinusoidal wave ac. The waveforms shown in Figure 11.24 are best observed by connecting an oscilloscope across the resistor.

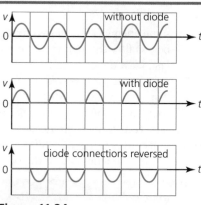

Figure 11.24

Diode bridges

■ Full-wave rectification (as shown in Figure 11.27) can be achieved by using four diodes in a *bridge circuit*.

> **Key concept**
>
> Half-wave rectification of an alternating current requires a single diode. Full-wave rectification needs four diodes in a 'bridge' circuit.

> **Expert tip**
>
> A typical **'bridge' circuit** contains two pairs of components connected in parallel, with a connection (a 'bridge') made between their intermediate points. Figure 11.25 shows an example using resistors (this circuit is called a Wheatstone bridge). The variable resistor is adjusted until no current flows through the meter, and then the value of an unknown resistance in the circuit can be determined very accurately. (This is *not* required in this course).
>
>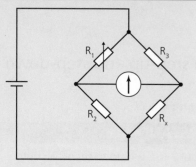
>
> **Figure 11.25**

■ Figure 11.26 shows four identical diodes connected in a **diode bridge circuit** across the output of a transformer. At a moment when point A is at a positive potential the current will follow the path ABCDEF. A little time later, as the output from the transformer changes, point F becomes positive and the current path is FBCDEA. The current through CD is always in the same direction.

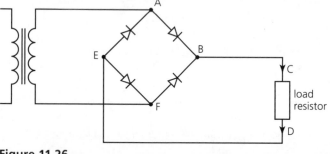

Figure 11.26

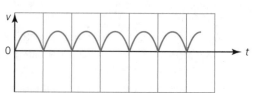

Figure 11.27

> **Expert tip**
>
> *Oscilloscopes* are used for displaying how a voltage varies with time. The input might be from a regularly repeating process (such as in Figure 11.24), or the more variable signals from the various types of sensor which provide a voltage output (a microphone, for example). The voltage scales can be easily adjusted for input p.d.s of most magnitudes and the timescale can be adjusted for variations over seconds or microseconds.

Investigating a diode bridge rectification circuit experimentally

■ An oscilloscope connected across the load resistor in Figure 11.26 will display the *full-wave rectification* variation of voltage shown in Figure 11.27. The effect of adding a capacitor can also be observed (see the next section).

Qualitatively describing the effect of adding a capacitor to a diode bridge rectification circuit

- Although the output from the bridge circuit shown in Figure 11.27 will result in a direct current through the load, it is certainly *not a steady* dc and will probably be unsuitable for many possible uses. The voltage can be *smoothed* by adding a capacitor across the output.
- The capacitor chosen must have a value so that if does not discharge too quickly or too slowly (no calculations needed).
- Figure 11.29 shows the effect of capacitor smoothing a full-wave rectified voltage. When the output from the rectifier circuit falls, the voltage across the capacitor falls more slowly and maintains the voltage across the load resistor. The rate at which the capacitor discharges depends on the values of the capacitor *and* the resistance (from Section 11.3: time constant, $\tau = CR$).

QUESTIONS TO CHECK UNDERSTANDING

25 Sketch the *V/I* characteristic of a diode to show the variation of current for both forward (+*V*) and reverse (−*V*) voltages.

26 a When an ac ammeter is used to measure the magnitude of an alternating current, does it display the peak value or the rms value?

b Explain why such an ammeter is unsuitable for measuring the current in a circuit such as that shown in Figure 11.26.

27 A 6 V volt (rms) supply was connected across a single diode and a lamp rated at 6.0 V, 3.0 W in series with each other.

a If the frequency of the supply was 100 Hz, sketch what would be seen on an oscilloscope connected across the lamp if it had a 10 cm wide screen and the beam producing the trace was moving from left to right across the screen at 5.0 m s⁻¹.

b What current would flow through the lamp if it was connected directly to the 6 V ac supply?

c Describe the appearance of the lamp when it was in series with the diode.

28 12 V dc power supplies are commonly used in school laboratories. Draw the circuit necessary for producing a smooth 12 V from a 230 V ac supply.

29 Consider Figure 11.29. Sketch the output from the same rectification circuit if the capacitor was replaced with another one of lower capacitance.

Key concept

If a suitable *capacitor* (see Section 11.3) is connected across the output of a rectifier, as in Figure 11.28, it can *smooth* the voltage so that there is less variation.

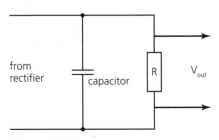

Figure 11.28

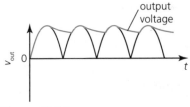

Figure 11.29

NATURE OF SCIENCE

Bias

Our experiences, education, friendships, culture, personality etc. all influence the way we process information and form opinions. This is inevitable, but it should be very much part of their *scientific method* that scientists are aware of their own (unintentional) *biases* and try to be objective about their own work and when reviewing the work of others. But, human nature being what it is, conflicts can still occur because of entrenched opinions and financial concerns. Towards the end of the nineteenth century there was extensive and sometimes irrational debate (especially in the USA) about whether ac or dc was better for electrical power transmission.

11.3 Capacitance

Revised ☐

Essential idea: Capacitors can be used to store electrical energy for later use.

Capacitance

Revised ☐

- The ability of any object to 'store' electric charge is described as **capacitance**. An electrical component designed to have significant capacitance is called a **capacitor**.
- Capacitors are useful components of circuits because of their ability to store small amounts of energy and the fact that the voltage across them can change in a predictable way with time.

Parallel plate capacitors

- When the switch in Figure 11.30 is connected to A, the battery will attract electrons off one plate and push them onto the other in a process we call *charging the capacitor*. As a plate becomes charged, electric forces will begin to resist further charging. For example, the extra electrons on the negative plate will tend to repel further electrons being added. Charging stops when p.d. across the plates has risen to become equal to the p.d. across the battery.

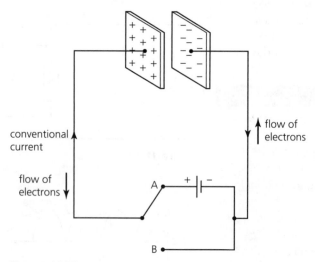

Figure 11.30

- When there is a p.d. across the plates, there will be a uniform electric field between them. This arrangement has been described in Chapters 5 and 10.
- Although the plates have equal and opposite charge, we sometimes say that the capacitor has *stored* charge.
- When the switch is connected to B the electrons reverse their movement and the *capacitor becomes discharged*. The discharge current transfers energy and therefore we can say that (electric potential) energy was stored on the charged capacitor.
- Charging and discharging occur very quickly unless there is significant resistance in the circuit.
- Figure 11.31 shows the circuit symbol for a capacitor. Because of their design, some capacitors must be connected the correct way around. They are marked with their polarity.

Factors affecting the charge on a plate in a parallel plate capacitor

- The charge, q, that builds up on each plate of a capacitor is proportional to (i) the p.d. between the plates, V, and (ii) the area of the plates, A. The arrangement is shown in Figure 11.32.
- The charge, q, on each plate of a capacitor is *inversely* proportional to the separation of the plates, d. (For example, doubling the separation will halve the charge, all other factors remaining unchanged.)
- The charge, q, on each plate of a capacitor also depends on the medium between the plates. We saw in Section 5.1 that the *electric permittivity* of a medium, ε, represents its electric properties.
- ε_0 is the permittivity of free space, which is similar in value to that for air. $\varepsilon_0 = 8.85 \times 10^{-12}\,\mathrm{C^2\,N^{-1}\,m^{-2}}$. All other materials have a permittivity greater than ε_0.
- Summarizing this information, the charge stored on a parallel plate capacitor can be calculated from the equation $q = \varepsilon \dfrac{VA}{d}$.

> **Key concept**
>
> Parallel metallic plates separated by an insulator are the most common kind of capacitor. When a p.d. is connected across the plates they will gain equal and opposite charge.

Common mistake

Note that the plates of a charged capacitor have equal and opposite charge: $+q$ and $-q$. The same charge has been moved from one plate to the other. When we refer to the charge on a capacitor, we refer to q, not $2q$, and the sign of the charge is not discussed.

Figure 11.31

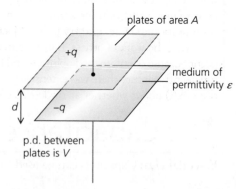

Figure 11.32

Calculating capacitance

- Capacitance is defined as the charge on either plate/potential difference between the plates. $C = \frac{q}{V}$. Capacitance has the SI unit of the **farad**, F. This is a large unit, so that µF and pF are in common use.
- Comparing the last two equations gives $C = \frac{\varepsilon A}{d}$, showing that a large value of capacitance requires plates of large area, very close together, with a medium of high electric permittivity between them.

> **Key concept**
>
> The capacitance of parallel plates increases if the plates have larger area and are closer together. Adding a suitable insulator (dielectric material), with high permittivity, between the plates also increases capacitance.

Dielectric materials

`Revised` ☐

- The insulating medium placed between the parallel plates of a capacitor is known as a **dielectric material**.

Describing the effect of different dielectric materials on capacitance

- Some charges within the molecules of a dielectric medium can move so that the molecules become **polarized**. Figure 11.33 shows that these polarized molecules align with the electric field. The overall effect is to reduce the electric field between the plates (compared to air). This allows more charge onto the plates for the same p.d., thereby increasing the capacitance, as shown in the equation highlighted above.

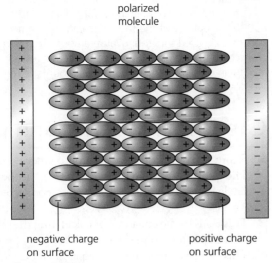

Figure 11.33

> **Expert tip**
>
> Larger values of capacitance generally require capacitors of increasing physical size (opposite to resistance). This means that keeping the plates very close together and using a dielectric material of high permittivity is important.
>
> The term *relative permittivity* is sometimes used to represent the ratio of the permittivity of a material to that of vacuum/air.

Solving problems involving parallel plate capacitors

QUESTIONS TO CHECK UNDERSTANDING

30 Sketch the electric field that exists between charged parallel plates.

31 Parallel metals plates, each of area 4.8 cm² are separated by 1.0 mm in air.

  a What charge will flow onto the plates when a p.d. of 80 V is connected across them?

  b The insulated plates are then disconnected from the voltage supply and their separation increased to 2.0 mm. What affect does this have on:

    i the charge on the plates

    ii the p.d. across the plates?

32 a What is the capacitance of a capacitor which gets charged to 4.5 × 10⁻⁶ C when connected across a p.d. of 1.5 V?

  b What total charge would be on the same capacitor if it was connected across 12 V?

33 A student makes a capacitor by placing 10 cm square pieces of aluminium foil on both sides of a thin sheet of plastic.

  a If the plastic was 0.12 mm thick and had a permittivity of 2.03 × 10⁻¹¹ C² N⁻¹ m⁻², what was the capacitance of the arrangement?

  b How much charge would flow onto this capacitor if it was connected to 6.0 V?

34 A certain kind of paper has an electric permittivity of 3.41 × 10⁻¹¹ C² N⁻¹ m⁻². If this paper was used in the space between the plates of a capacitor instead of air, by what factor would the capacitance increase?

Determining the energy stored in a charged capacitor

- Figure 11.34 shows how the charge on a capacitor changed as the p.d. across its plates is increased.
- The gradient of this graph is equal to the capacitance ($C = \frac{q}{V}$).
- Since V = energy/charge, the energy stored on a capacitor, E = average value of charging voltage multiplied by charge, $E = \frac{1}{2}qV$. This also equals the area under a V–q graph like that shown in Figure 11.34.
- Or, since $q = CV$, $E = \frac{1}{2}CV^2$.

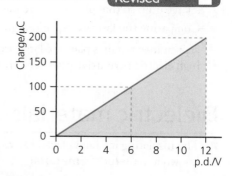

Figure 11.34

> ### Expert tip
>
> A 1 F capacitor is a relatively large value capacitance and it might be the size of a cup. When it is connected to, for example, 100 V, the energy stored is only 5000 J. A battery of the same size can store much more energy, so capacitors are not usually used for storing large amounts of energy. However, capacitors may be considered to have no effective internal resistance so that they can be very useful in delivering (relatively small amounts of) energy very quickly, for example to power the flash of a camera.

> ### Key concept
>
> Capacitors can store small amounts of energy ($\frac{1}{2}CV^2$) which can then be delivered to a circuit at a rate depending upon the resistance in the circuit.

Capacitors in series and parallel

- Figure 11.35 shows three capacitors connected in parallel across the same p.d. The total charge stored, q ($= q_1 + q_2 + q_3$), is the same as would be stored on a single capacitor of capacitance $C_{parallel}$ connected to the same voltage:
$q = VC_{parallel} = VC_1 + VC_2 + VC_3$.
- This leads to the equation for calculating the overall capacitance of two or more capacitors connected in *parallel*: $C_{parallel} = C_1 + C_2 + \dots$
- This can be compared to the equation for calculating the overall capacitance of two or more capacitors connected in *series*: $\frac{1}{C_{series}} = \frac{1}{C_1} + \frac{1}{C_2} + \dots$

- When capacitors are connected in series the charge which flows off one must equal the charge which flows onto another. That is, they must all have the same charge. Using this fact together with $V = V_1 + V_2 + V_3$ leads to the previous highlighted equation.

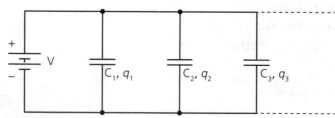

Figure 11.35

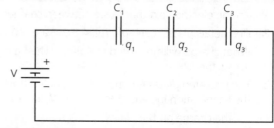

Figure 11.36

▣ Investigating combinations of capacitors in series and parallel

- The value of an unknown, uncharged capacitor (or combination of capacitors) can be found by connecting it in parallel across a known, charged capacitance and observing the change in p.d. See Figure 11.37. When the switch is closed the capacitors will share the charge and the value of V will fall.
- For example, consider a capacitor, C_1, of capacitance 480 μF charged by 12.0 V connected to an uncharged capacitor, C_2, of unknown capacitance. The total charge, q, can be assumed to be constant, but it is now shared. If the voltage across the combination fell to 8.0 V, $q = C_1V_1 = C_{comb}V_{comb}$. Hence the capacitance of the combination can be determined (720 μF) and the equation for capacitors in parallel used to determine the unknown capacitance (240 μF).
- Alternatively an unknown capacitance can be determined from the rate of decrease of current (or voltage) when it discharges through a known resistance (see below).

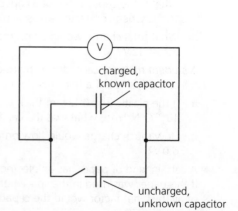

Figure 11.37

QUESTIONS TO CHECK UNDERSTANDING

35 Consider Figure 11.34. Determine:

  a the value of the capacitor represented

  b the energy stored when the p.d. across the capacitor was 6.0 V.

36 The voltage used in a defibrillator was 1000 V. See Figure 11.38. What capacitor value would be able to store energy of 500 J to supply a patient at this voltage?

37 The flash of a camera is powered by a 180 μF capacitor. Estimate the average power of a 250 V flash that had a duration of 1/200 second.

38 a What four values of combined capacitance can be made by connecting three 27 pf capacitors together?

  b Describe how you could confirm experimentally that your answers were correct.

39 A 22 μF capacitor charged to 9.0 V was connected across an uncharged 33 μF capacitor.

  a What was the voltage across them after the connection was made?

  b Calculate the energy stored:

    i   on the original charged capacitor

    ii on the combination.

  c Account for the difference between these energies.

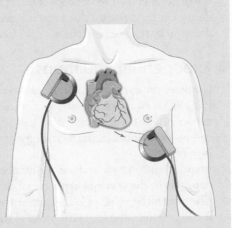

Figure 11.38

Resistor–capacitor (RC) circuits

Revised

- Unless it is required to discharge a capacitor very quickly, a resistor will be included in any capacitor circuit in order to control the rate of discharging (or charging).
- The charging or discharging of a capacitor can be observed and recorded by measuring either the changing p.d. across it, or the changing current through the resistor (no current actually flows *through* the capacitor). See Figure 11.39. which shows a **resistor-capacitor (RC) circuit** which can be alternately charged and discharged. When the switch is moved to A the capacitor will charge through the resistor, and it will discharge through R when connected to B. If required, a small, low voltage capacitor can be discharged quickly by putting a connecting lead across its terminals.

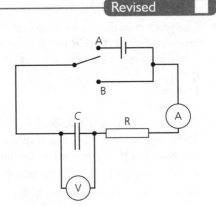

Figure 11.39

▌ Describing the nature of the exponential discharge of a capacitor

- When a capacitor discharges, the rate of discharge will be highest at the beginning because the amount of charge on the plates is greatest, so that the free electrons experience the greatest forces. The rate of discharge decreases with time, but theoretically never reaches zero.
- Figure 11.40 shows a typical discharge graph. Note that it may represent the variation with time of the discharge current, or the voltage across the capacitor, or the charge remaining on the capacitor plates.
- The graph is characterized by the fact that the rate of discharge at any time (which could be determined from the gradient of the graph at that moment) is proportional to the value of the quantity at that moment. For example, $-\frac{\Delta Q}{\Delta t} \propto Q$. This is the necessary condition for an **exponential decrease**.
- In equal time intervals the values shown on the graph decrease to the same fraction. That is, the ratio of the values at the beginning and end of *any* equal time intervals are always the same. For the example of Figure 11.40, the value of each (equally spaced) point is 67% of the previous point.

> **Key concept**
>
> During a **capacitor discharge**, the charge, voltage and current decrease exponentially. That is, their values fall by the same fraction in equal time intervals.

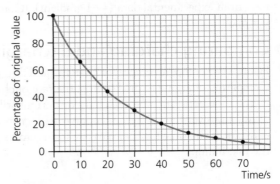

Figure 11.40

■ As with the exponential decay of the count from a radioactive source (Section 7.1), we cannot quote a time for the process to end (because in theory it will *never* end). The concept of *half-life* is used with irreversible radioactive decay, but for capacitor discharge it is preferred to characterize the exponential decrease by the use of a *time constant* (see below).

■ Exponential equations

■ Consider the exponential decrease shown in Figure 11.41. If any quantity has a value N_0 at the beginning of a time interval t and then decreases exponentially (as described above), its value at the end of the time interval, N, can be represented by the following equation:

$$N = N_0 e^{-kt}$$

where k is a constant and e is the number 2.718. This number arises naturally in the st udy of exponential changes.

■ The *greater* the value of k, the *quicker* the decrease.

■ For RC circuits, the *larger* the value of the capacitor and/or the resistor, the *slower* the rate of discharge, and $k = \frac{1}{RC}$. We can now write down three similar equations to represent the exponential decrease of the three quantities during discharge (there is no requirement to understand the origin of these equations):

$$q = q_0 e^{-\frac{t}{RC}}$$
$$I = I_0 e^{-\frac{t}{RC}}$$
$$V = V_0 e^{-\frac{t}{RC}}$$

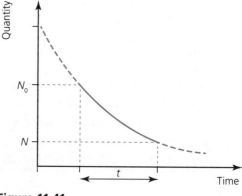

Figure 11.41

■ Time constant

■ The product of the capacitance and resistance of a circuit is called the circuit's time constant, τ. $\tau = CR$ (unit: seconds, s).

■ The three equations above can therefore be written as:

$$q = q_0 e^{-\frac{t}{\tau}}$$
$$I = I_0 e^{-\frac{t}{\tau}}$$
$$V = V_0 e^{-\frac{t}{\tau}}$$

> ### Key concept
> The **time constant**, τ, is equal to the time that it takes for the charge (or current or voltage) to decrease to $1/e \times$ the original value (37%).

■ Consider a numerical example for a circuit containing a capacitor of $78\,\mu F$ and a resistor of $560\,k\Omega$. The time constant of this circuit is $44\,s$. If the capacitor was charged to $12\,V$ (V_0) before the discharging began, the charge on the capacitor would then be $9.4 \times 10^{-4}\,C$, the initial current (I_0) would be $21\,\mu A$. After $44\,s$ all three of these values will have fallen to 37% of their initial values.

■ If we want to find values at other times, we need to use one, or more, of the equations highlighted above. For example, suppose we want to know the current, I, through the resistor after $10\,s$: $I = I_0 e^{-\frac{t}{\tau}} = (21 \times 10^{-6}) \times e^{-\frac{10}{44}}$; taking natural logarithms: $\ln I = \ln(21 \times 10^{-6}) - 0.227$; leading to $I = 17\,\mu A$.

■ If we take natural logarithms of the equations highlighted above we get (using current as an example): $\ln I = \ln I_0 - \frac{t}{\tau}$.

■ A graph of $\ln I$–t will be a straight line with a negative gradient, as shown in Figure 11.42. (The values of $\ln I$ will probably be negative because the currents are usually small.)

■ A *straight-line* graph like this provides the best way of determining a value for τ from experimental data ($= -1/\text{gradient}$). (Compare $\ln I = -\frac{t}{\tau} + \ln I_0$ to $y = mx + c$.)

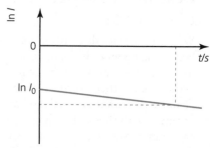

Figure 11.42

■ Comparing charging and discharging graphs

■ Figure 11.43 compares graphs for charging and discharging.

■ Comparing charging to discharging, the rate of charging is still controlled by the time constant, and the charging current still falls by 37% in a time of RC, however the voltage and charge are *increasing*, so that the time constant is the time for V or q to rise from zero to $(100 - 37)\% = 63\%$ of the maximum.

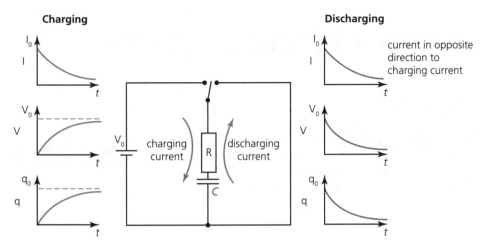

Figure 11.43

■ Solving problems involving the discharge of a capacitor through a resistor / Solving problems involving the time constant of an RC circuit for charge, voltage and current

QUESTIONS TO CHECK UNDERSTANDING

40 Show that the units of the time constant are seconds.

41 An unknown capacitor was discharged through a 11 kΩ resistor and the current in the circuit fell to 37% of its initial value of 54 mA in a time of 12 s.

a What was the value of the capacitance?

b What was the value of the current after another 12 s?

42 A 6.0 V battery was used to fully charge a 3.9 μF capacitor. It was then discharged through a 68 kΩ resistor.

a What was the initial discharge current?

b Calculate the time constant of the circuit.

c Sketch a graph to show how the voltage across the capacitor changed during the first second of discharge.

43 A student recorded the discharge of a capacitor by measuring the p.d. across it at regular times. After two minutes the voltage was 4.69 V and one minute later it had fallen to 3.22 V.

a Determine the time constant of the circuit.

b Calculate the voltage across the capacitor after a further five minutes.

44 A fully charged 24 V, 68 μF capacitor is discharged through a 5.6 kΩ resistor.

a Determine the time constant of the circuit.

b How much charge was initially stored on the capacitor?

c Write down an exponential equation to represent how the charge on this capacitor changed with time.

d Calculate the time interval until the charge on the capacitor has fallen to 4.5×10^{-5} C.

45 A 12 V battery is connected across an uncharged 2.7 μF capacitor and an 8.2 MΩ resistor in series. Sketch two graphs (on the same axes) to show how the voltages across the capacitor and resistor change during the first 70 s.

46 Consider again Figure 11.43, which represents the discharge in an RC circuit. The value of $\ln I_0$ was −9.57 and $\ln I$ was −12.14 at time 100 s.

a What was the initial discharge current?

b Determine the time constant of the circuit.

c If the resistance in the circuit was 820 kΩ, what was the capacitance of the circuit?

NATURE OF SCIENCE

■ Relationships

There are many examples of exponential changes to be found in physics. In particular, in this course we use the mathematics of exponentials in the study of capacitors and radioactive decay, but the same branch of mathematics has many applications beyond physics and science.

12.1 The interaction of matter with radiation

Essential idea: the microscopic quantum world offers a range of phenomena, the interpretation and explanation of which require new ideas and concepts not found in the classical world.

Photons

- ■ In Chapter 7 we saw that electromagnetic radiation can be considered to be composed of large numbers of small 'packets' of energy called *photons* (or *quanta*) and that the energy of each photon, E, can be determined from $E = hf$, where h represents *Planck's constant* and f represents the frequency of the radiation.
- ■ Photons are emitted or absorbed when there are changes in the energy levels of atoms, molecules and ions.
- ■ However, the most convincing evidence for the quantum (photon) theory of radiation is provided by the *photoelectric effect*.

The photoelectric effect

- ■ We know from Section 5.1 that metals contain *free electrons*. These electrons may be emitted from some clean metal surfaces when the metal is illuminated by suitable electromagnetic radiation, usually ultraviolet radiation (the photons of which carry more energy than visible light). This phenomenon is called *photoelectric emission*, or the **photoelectric effect**.

- ■ **Discussing the photoelectric effect experiment and explaining which features of the experiment cannot be explained by the classical wave theory of light**

- ■ The photoelectric effect can be demonstrated using an ultraviolet source, a clean zinc plate and a very sensitive charge measuring instrument (a coulombmeter). The principle is shown in Figure 12.1. The emitted electrons are called **photoelectrons**.

> **Key concept**
>
> For a particular metal surface, photoelectrons are only emitted for radiation which has a frequency above the metal's *threshold frequency*. The intensity of the radiation does not affect the energies of individual photoelectrons.

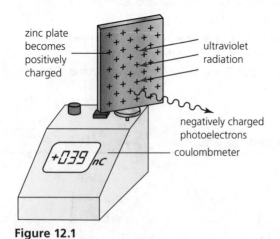

zinc plate becomes positively charged

ultraviolet radiation

negatively charged photoelectrons

coulombmeter

+039 nC

Figure 12.1

> **Expert tip**
>
> In practice, the photoelectrons in Figure 12.1 will be attracted back to the plate and it is better to first charge the plate negatively, so that the photoelectrons are repelled, and the magnitude of the negative charge on the plate decreases.
>
> Figure 12.2 shows an alternative demonstration using a gold-leaf electroscope.

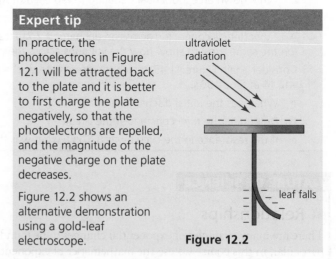

ultraviolet radiation

leaf falls

Figure 12.2

- The most important observations from these and related experiments on the photoelectric effect (see below) are:
  - □ for photoelectric emission to occur, there is a **threshold frequency**, f_0, below which *no* electrons are released (regardless of the intensity or the duration of the radiation)
  - □ there is never any measurable time delay between the radiation striking the surface and the emission of the photoelectrons (regardless of the intensity of the radiation)
  - □ the maximum energy of the emitted electrons depends on the *frequency* of the radiation, but *not* the intensity
  - □ increasing the intensity of the radiation increases only the *number* of photoelectrons emitted.
- None of these observations can be explained by the *classical wave theory of light*. Using the wave theory we might predict (wrongly) that the effect would occur with radiation of any frequency, and that the photoelectrons would be emitted only after sufficient time had passed for the waves to deliver enough energy to the surface of the metal, and that greater intensity radiation would produce more energetic photoelectrons.
- A minimum (threshold) frequency is observed because individual photons require a minimum energy for the photoelectric effect to occur with a particular metal ($E = hf$).

> **Key concepts**
>
> Einstein's explanation of the photoelectric effect assumes that *individual photons* (not waves) each transfer all of their energy to individual electrons and, *only if that energy is enough*, will the electrons be able to escape from the surface.
>
> The *work function*, Φ, of a metal is the *minimum* energy required to remove a photoelectron from its surface.

■ Einstein's equation and the work function

- We will now discuss the photoelectric effect in more mathematical detail. We must first identify the **work function**, Φ, of a metal surface, which is the *minimum* energy needed to free an electron from that surface. The electrons need kinetic energy to overcome the forces from positive charges pulling them back to the metal.
- The units usually used for work function are *electron volts*, eV (see Section 5.1). Different metals have different work functions. For example, the work function of zinc (as mentioned above) is 4.33 eV (= 6.94×10^{-19} J). This means that for electromagnetic radiation to produce the photoelectric with zinc, its photons must have *at least* this energy. Using $E = hf$ shows that the minimum frequency needed is 1.0×10^{15} Hz, which is in the ultraviolet part of the spectrum. Visible light photons do not have enough energy to cause photoelectric emission with zinc.
- It is important to realize that electrons in different locations will require different amounts of energy to escape from the surface. The *work function* is just the *minimum* amount required for a particular metal.
- If the photoelectric effect occurs at a particular frequency for a particular metal, then the energy carried by the photons = work function + *maximum* energy of photoelectrons, or in symbols:
- $hf = \Phi + E_{max}$. This is called the **Einstein photoelectric equation**, which may also be written as: $E_{max} = hf - \Phi$
- Figure 12.3 represents this equation graphically. Comparing this to equations of the form $y = mx + c$, we see that the magnitude of the intercept is the work function, Φ, and the gradient is Planck's constant, h.
- The threshold frequency, f_0, occurs when $E_{max} = 0$. That is, $hf_0 = \Phi$.
- Because Planck's constant is a *fundamental* constant, graphs for all metals will have the same gradient.

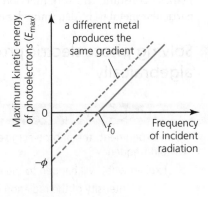

Figure 12.3

■ Experiment to determine Planck's constant and work function

- If we wish to use a photoelectric emission experiment to determine values of Planck's constant and the work function of the metal used, we must develop a way of measuring E_{max}. Figure 12.4 shows a suitable arrangement.

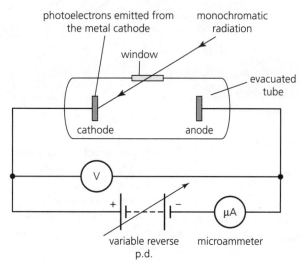

Figure 12.4

Key concept

The maximum energy of emitted photoelectrons can be determined by increasing the size of a 'reverse' voltage until they are all repelled.

- Photoelectrons are emitted from the metal surface of the cathode when suitable electromagnetic radiation is incident upon it. The frequency of the radiation is needed in the calculations and radiation of a single frequency is desirable (*monochromatic*). Because there is no air in the tube (it is *evacuated*), the photoelectrons are able to travel to the anode without losing any kinetic energy.
- The movement of the negatively charged photoelectrons is controlled by the p.d. across the tube. If the anode is made positive, electrons will be attracted to it, but if a **reverse voltage p.d.** is applied, the electrons will be repelled.
- The reverse p.d. is slowly increased from zero until even the most energetic photoelectrons (those with energy E_{max}) just cannot reach the anode. At this time the current will just be reduced to zero. This is shown in Figure 12.5.
- The maximum kinetic energy lost by photoelectrons = electric potential energy gained by those electrons: $E_{max} = eV_s$, where V_s is the reverse voltage which just stops the photoelectric current.
- We can now re-write the Einstein equation ($E_{max} = hf - \Phi$) in terms of a quantity we can measure directly, V_s:

$$eV_s = hf - \Phi$$

- Measurement of the stopping voltage, V_s, for different frequencies enables Planck's constant, the work function and threshold frequency for any particular metal to be determined. See Figure 12.6 for an example.

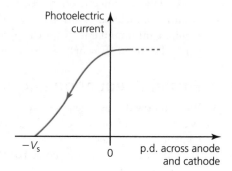

Figure 12.5

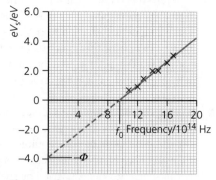

Figure 12.6

■ Solving photoelectric problems both graphically and algebraically

QUESTIONS TO CHECK UNDERSTANDING

1 In an experiment, the photoelectric effect is detected from a clean metal surface with electromagnetic radiation of a certain frequency.

   a Explain what will happen to the *energy* of the photoelectrons and the *number* of photoelectrons when:

   i the intensity of the radiation is increased (at the same frequency)

   ii the frequency of the radiation is increased (at the same intensity).

   b With the original intensity and frequency, the metal is changed for another with a different work function. Explain why photoelectric emission may stop.

2 Sketch a labelled E_{max}–f graph for photoelectric emission from magnesium ($\Phi = 3.7\,eV$)

3 The work functions of caesium, iron and platinum are 2.1 eV, 4.5 eV and 6.4 eV.

   a Which metal is the most likely to release photoelectrons with visible light?

   b Convert 4.5 eV to joules.

   c What is the threshold frequency for iron?

   d What is the threshold wavelength for platinum?

   e What is the maximum energy of photoelectrons emitted from iron by radiation of frequency $8.4 \times 10^{15}\,Hz$?

4 Consider Figure 12.6. Using the graph, determine values for:

   a the threshold frequency

   b the work function

   c Planck's constant.

5 In an experiment such as that shown in Figure 12.4, calculate the value of the reverse p.d., V_s, which would be needed to just prevent a photoelectric current with radiation of frequency 1.2×10^{15} Hz if the metal had a work function of 2.3 eV.

6 Make a copy of Figure 12.7, which shows the variation of photoelectric current with p.d. for a certain metal and radiation of a fixed frequency and intensity. On the same axes, draw a second graph to show what would happen if the intensity of the radiation was increased.

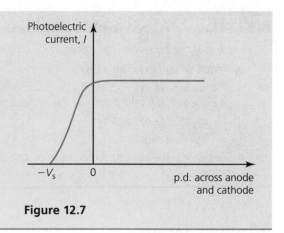

Figure 12.7

Quantization of angular momentum and the Bohr model for hydrogen

■ In Section 7.1 we saw that the spectral analysis of the light emitted from excited atoms of hydrogen shows that the atoms have a series of discrete energy levels. These are shown in Figure 12.8. The atoms of other elements have similar (but more complicated) energy levels.

■ Niels Bohr was the first to develop a numerical description of the discrete **energy levels of hydrogen atoms**.

■ For hydrogen, the simplest atom, the energy of these levels had been found to fit the *empirical* mathematical formula: $E = \frac{-13.6}{n^2}$ eV, where n is an integer, known as the *principal quantum number*. Remember that the energy levels have *negative* values because energy has to be supplied to free the electron from the attraction of the nucleus.

■ The Bohr model of the hydrogen atom explains these levels by stating that electrons orbit the nuclei at a constant radius because of the electric force of attraction, *but* the electrons could only be in certain precise orbits such that their angular momentum (mvr) could only have *discrete* values according to the equation $mvr = \frac{nh}{2\pi}$. (n is an integer, m is the mass of the electron and v its speed in an orbit of radius r.) See below for an explanation.

■ This equation very successfully predicted the energy levels of the hydrogen atom.

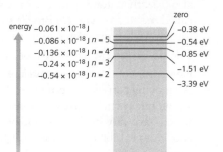

Figure 12.8

Key concept

In the Bohr model of the hydrogen atoms the electron was in constant circular orbit around the proton, but its orbital radius could only have one of a series of possible values, dependent upon its angular momentum.

Expert tips

Classical physics predicts that an orbiting charged particle (an electron) will emit electromagnetic radiation and spiral inwards as its energy decreases. The Bohr model contradicted this concept.

The *rotational angular momentum* of any mass moving in a circular path is equal to the linear momentum, mv, multiplied by the radius, r, of the motion. The SI unit is $kg\,m^2\,s^{-1}$. Angular momentum is covered in Option B.

■ With the discovery that electrons have wave properties (see the next section), Bohr's equation could be explained by considering that a whole number of electron wavelengths had to fit into a circumference of a circle around the nucleus (see Figure 12.10). The possible values for the discrete wavelengths are then $\frac{2\pi r}{1}, \frac{2\pi r}{2}, \frac{2\pi r}{3}$, etc. That is, possible values for $\lambda = \frac{2\pi r}{n}$. But we know the de Broglie wavelength of the electron (see below), $\lambda = \frac{h}{p}$ or $\frac{h}{mv}$. Combining these wavelength equations leads to $mvr = \frac{nh}{2\pi}$, as above.

■ This theory for understanding the energy levels of hydrogen does not apply so easily to more massive atoms with many electrons, but the basic principle of explaining energy levels using standing waves still applies.

QUESTIONS TO CHECK UNDERSTANDING

7   Show that the equation $\frac{-13.6}{n^2}$ successfully predicts the energy levels of hydrogen shown in Figure 12.8.

8   a   If the electron in a hydrogen atom in its ground state is orbiting at a distance of 5.3×10^{-11} m from the nucleus, calculate:

   i   its angular momentum

   ii   its linear momentum.

   b   Determine the angular momentum of an electron in the lowest excited state.

Matter waves

Revised ▢

- Electromagnetic radiation has some properties that require an explanation in terms of waves (diffraction and interference), and some properties (photoelectric emission) which require a photon/particle explanation. This is known as **wave–particle duality**.

- De Broglie was the first to expand this concept to suggest the 'opposite' was also true: that atomic particles have wave properties, so that they may be called **matter waves**. De Broglie suggested that the wavelength of a particle was inversely proportional to its momentum.

- The **de Broglie wavelength** of a particle, $\lambda = \frac{h}{p}$, where p represents the momentum of the electron and h represents Planck's constant.

- All moving particles, in principle, show wave-like properties. However, only sub-atomic particles (like electrons, protons and neutrons) and atoms or ions which have very small mass (and momentum), will have a wavelength large enough to exhibit observable wave properties (diffraction and interference) in experiments.

> **Key concept**
>
> **The de Broglie hypothesis:** all particles have wave properties, each with a wavelength inversely proportional to its momentum. The de Broglie wavelength, $\lambda = \frac{h}{p}$

■ Discussing experimental evidence for matter waves, including an experiment in which the wave nature of electrons is evident

- The wave nature (diffraction) of electrons can be demonstrated in a school laboratory with equipment like that shown in Figure 12.9.

> **Key concept**
>
> The diffraction of an electron beam as it passes though the regular structure of a thin (graphite) crystal is evidence for the wave nature of the electrons.

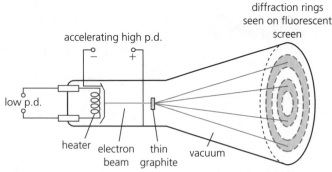

Figure 12.9

- The electrons are accelerated to high speeds so that they have enough energy to pass through a very thin graphite crystal, which *diffracts* the electron waves in a way similar to a diffraction grating diffracting light (Section 9.3). Diffraction *rings* (rather than fringes or spots) are seen because of the random arrangement of crystalline areas within the graphite.

- As with the diffraction of light through a grating, the measurement of diffraction angles can be used with a knowledge of the spacings of the diffracting structure (separation of carbon atoms) to determine the wavelength. The two rings in Figure 12.9 are caused by diffraction from two particular sets of layers of atoms within the graphite crystal.

- If an electron of mass m_e and charge e is accelerated by a potential difference of V, then we can use $E_K = \frac{p^2}{2m_e}$ (from Section 2.3) $= eV$, and $\lambda = \frac{h}{p}$ to determine the de Broglie wavelength of the electron.

Expert tip

The Davisson–Germer experiment was the first to confirm the de Broglie hypothesis. Similar in principle to the experiment shown in Figure 12.9, it involved scattering a beam of electrons from a nickel crystal. The electrons were diffracted by the regularity of the arrangement of atoms in the surface of the nickel and constructively interfered only at certain angles.

■ Fitting electron waves into atoms

- The discovery that electrons had wave properties opened the way to understanding why atoms have energy levels.
- In Section 4.4 we learned that confined waves can exist only as a series of *standing waves* with different wavelengths. Electrons exist within atoms with discrete energies because they are standing waves of discrete wavelengths.
- Figure 12.10 shows simplified examples: fitting three, four and five wavelengths into the circumference of an orbit. Each of these would correspond to a different discrete wavelength and energy for the electron.
- For many reasons this is a much simplified model, but it does illustrate an important principle: how an appreciation of electron standing waves leads to an understanding of the quantization of energy levels.

Key concept

Electrons within atoms exist as standing waves. These waves can only have certain discrete wavelengths and energies.

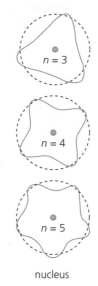

nucleus

Figure 12.10

QUESTIONS TO CHECK UNDERSTANDING

9 What is the wavelength associated with an alpha particle travelling at 5% of the speed of light? (Assume mass of alpha particle $= 6.64 \times 10^{-27}$ kg.)

10 What is the approximate speed of an electron that has the same wavelength as a typical X-ray?

11 In an experiment like that shown in Figure 12.9 electrons were accelerated by a p.d. of 5000 V.

   a Determine the speed of the accelerated electrons (assume they started from rest).

   b Calculate the momentum of an accelerated electron.

   c What is the de Broglie wavelength of the electrons?

   d If the electrons were diffracted by layers of carbon atoms of spacing 1.4×10^{-10} m, use the diffraction grating equation to estimate the angle through which they might be diffracted.

   e Discuss the effect on all your answers if the p.d. accelerating the electrons was doubled.

Pair production

Revised ☐

- Another example of the interaction of matter with radiation is the production of a pair of particles when a (gamma ray) photon interacts with a nucleus. The photon is converted into two particles, the nucleus is unchanged. This is only possible if the two particles are a particle–antiparticle pair (so that the conservation rules apply). This is called **pair production**.
- Antimatter was briefly introduced in Section 7.3. For every elementary particle there is an antiparticle which has same mass as the particle, but with the opposite charge (if it is charged), and opposite quantum numbers. A few uncharged particles, like photons, are their own antiparticles.
- An energetic gamma ray photon can undergo pair production ($\gamma \rightarrow e^- + e^+$) if it passes into the strong electric field close to a nucleus. The reason why a nucleus is needed is explained below.

Key concept

An electron–positron pair can be produced from a gamma ray if it has enough energy and interacts with a nucleus.

■ Figure 12.11 represents the production of an electron–positron pair. Figure 12.12 shows a Feynman diagram of the interaction.

■ A more energetic photon may be able to produce pairs of more massive particles, for example, a muon and antimuon.

■ In order for pair production to be possible, the energy carried by the photon must be greater than the sum of the rest mass-energies of the particles produced. For an electron–positron pair the total rest mass-energy is $2 \times 0.511\,\text{MeV}c^{-2}$. This corresponds to a minimum photon frequency of $2.47 \times 10^{20}\,\text{Hz}$. If the photon has higher frequency (greater energy) the extra energy will be given to the kinetic energy of the electron and positron.

■ A third particle, the nucleus, is involved because if only two particles were involved momentum could not be conserved. The nucleus will change momentum, but is otherwise unaffected.

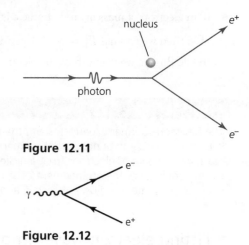

Figure 12.11

Pair annihilation

■ Figure 12.13 shows the annihilation of an electron–positron pair $(e^- + e^+ \rightarrow \gamma + \gamma)$, and Figure 12.14 represents the same event with a Feynman diagram.

■ If we assume that the particles had insignificant combined momentum, then the two gamma rays must move off in opposite directions (conservation of momentum), each with half the combined mass-energy of the original particles.

> **Key concept**
>
> When a particle interacts with its antiparticle their mass-energy is completely transferred into two gamma ray photons (or, at higher energies, other particles). This is called **pair annihilation**.

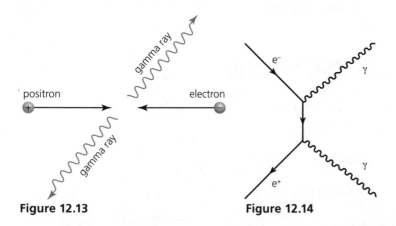

Figure 12.13   **Figure 12.14**

QUESTIONS TO CHECK UNDERSTANDING

12 Explain why a third particle (a nucleus) is necessary in pair production..

13 a Show why a minimum photon frequency of $2.47 \times 10^{20}\,\text{Hz}$ is needed for pair production.

  b What is the wavelength of this gamma ray?

14 Explain why you would not normally expect pair annihilation to result in a single photon.

Probability waves

■ Diffraction and interference patterns can be explained satisfactorily using the *wave model* for electrons and electromagnetic radiation, but when we try to understand these phenomena using the *photon model* we immediately have problems: for example, how is it possible that identical photons following the same path under identical conditions can move in different (non-random) directions after passing through a slit?

> **Key concept**
>
> The exact behaviour of individual photons/particles is unpredictable. This *uncertainty* is the true nature of the sub-atomic (quantum) world, it is *not* a statement that we do not have enough information, or that it is too difficult to know.

- However, it *is* possible to predict the overall behaviour of large numbers of photons/particles. For example we may know with 'certainty' the shape of a diffraction pattern, but not the behaviour of the individual photons that produced it. Figure 12.15 shows a representation of the distribution of photons which have arrived in a diffraction pattern at *very* low intensity. The developing pattern is seen but the locations of the arrival of individual photons was unpredictable, and only become known *after* they had arrived.

- Predicting the exact individual behaviour of particles is not possible, we can only describe the probabilities of a range of possible outcomes. This uncertain behaviour of photons/particles (including electrons) is fundamental in atomic physics and it was very important that a mathematical description of the probabilities was developed. This came with the *wave function* developed by Schrödinger and others:

Figure 12.15

The wave function

- In the macroscopic world we would describe the motion of a ball (for example) by stating its position and velocity, but in the sub-atomic world, these properties of a particle are always uncertain. A description in terms of probabilities is needed.

- The *wave function of a particle*, is a mathematical model involving probabilities and there is no simple visualization. Further mathematical details are *not* needed in the IB Physics course.

- The wave function may be considered to be an equation describing the amplitude of the particle wave (de Broglie wave), but there is no physical significance to that interpretation.

- The wave function of a particle extends throughout all space. This means that, theoretically, a particle can be located anywhere and everywhere, although most probabilities are vanishingly small.

- The square of the wave function has very important significance. The probability of finding a particle at any point in space is predicted by the square of the absolute amplitude of its wave function at that point, $|\Psi|^2$.

> **Key concepts**
>
> The **wave function** of a particle, $\Psi(x, t)$, fully describes the state of a particle at any given point in space at any given time.
>
> The probability of finding a particle at a distance r from a reference point, in unit volume, V, is known as its **probability density**, $P(r)$. $P(r) = |\Psi|^2 \Delta V$

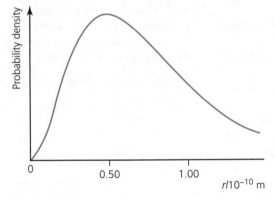

Figure 12.16

- Figure 12.16 shows a simple example, the probability density of an electron in a hydrogen atom. We cannot know for sure where the electron is. It is everywhere (except in the nucleus), with varying probabilities: the graph shows that the electron is most likely to be at a distance of 0.5×10^{-10} m from the nucleus.

QUESTIONS TO CHECK UNDERSTANDING

15 Explain why a wave function is needed to describe an electron.

16 Scientists sometimes refer to 'electron clouds' and Figure 12.17 shows a *cloud* which corresponds approximately to Figure 12.16. Discuss the use of this term.

17 Figure 12.18 shows the probability density for electrons in an atom and how it varies with distance from the nucleus. Describe where the electrons are most likely to be located.

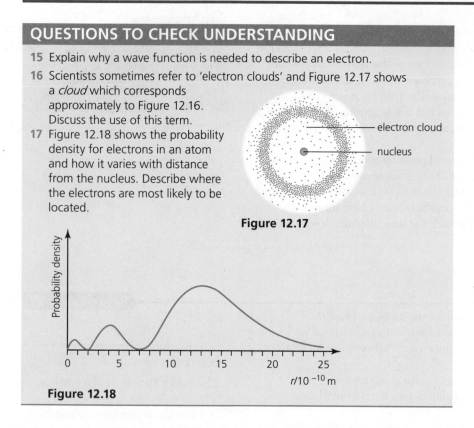

Figure 12.17

Figure 12.18

The uncertainty principle for energy and time, and position and momentum

- The position, x, and momentum, p, of a moving photon/particle cannot be known exactly. The **Heisenberg uncertainty principle** quantifies this uncertainty by linking two variables: $\Delta x \Delta p \geq \frac{h}{4\pi}$. Δx represents the uncertainty in the measurement of position and Δp represents the uncertainty in the measurement of momentum. Position and momentum are described as **conjugate variables**.

- From this equation we can see that the *more* precisely the position of a photon/particle is specified (smaller Δx) the *less* precisely the momentum is known (greater Δp), and vice versa. If one quantity is known perfectly, the other has infinite uncertainty.

- If a particle has a well-defined de Broglie wavelength, then its momentum is known precisely, so there can be no knowledge of its position.

- Energy, E, and time, t, are two other variables linked by the uncertainty principle: $\Delta E \Delta t \geq \frac{h}{4\pi}$.

- The consequences of the uncertainty principle are considerable, as the following examples show.

> **Key concepts**
>
> The equations expressing the *Heisenberg uncertainty principle* show that the more precisely the position of a particle is known, the less certainty there is about its momentum (and vice versa).
>
> Energy and time are similarly interconnected.

■ Stating order of magnitude estimates from the uncertainty principle

■ I Energy of the ground state of an atom

- We can use an estimate of the radius of a hydrogen atom to determine an approximate value for the minimum *kinetic energy*, E_K, of an electron in that atom: if radius, $r \approx 0.5 \times 10^{-10}$ m and we assume that the position of the electron is uncertain within that distance, then $\Delta x = r = 0.5 \times 10^{-10}$ m, and Δp can be determined from $\Delta x \Delta p \geq \frac{h}{4\pi}$. Then $E_K = \frac{p^2}{2m}$.

- If required, the electric *potential energy* can be determined from $E_P = \frac{kq_1q_2}{r}$ (Chapter 10). If we imagine that an atom was smaller, then the kinetic energy of the electron (as calculated above) would increase while the potential energy would decrease. The actual size of the atom ($r \approx 0.5 \times 10^{-10}$ m) is such that it results in the smallest *total* energy (kinetic energy + potential energy).

■ II Electrons cannot exist in nuclei

■ If we propose that an electron is positioned somewhere in the nucleus of a hydrogen atom of diameter $\approx 10^{-15}$ m, then 10^{-15} m is the uncertainty in its position (Δx). Inserting this value in $\Delta x \Delta p \geq \frac{h}{4\pi}$ leads to $\Delta p \approx 5 \times 10^{-20}$ kg m s^{-1}. An electron with this momentum would have too much energy to be retained within the hydrogen nucleus. So, the uncertainty principle explains why electrons cannot be found in nuclei.

■ III Lifetime of an electron in an excited state

■ Suppose an electron jumps from the ground state of an atom to an excited energy state which is 8 eV higher. If it remains in that state for 2×10^{-9} s, then the uncertainty in the energy level (from $\Delta E \Delta t \geq \frac{h}{4\pi}$) is $\approx 10^{-7}$ eV. This is very small percentage of the energy level, which means that it can be determined accurately.

QUESTIONS TO CHECK UNDERSTANDING

18 Figure 12.19 shows the wave function of two electrons at a certain time.

   a Which electron has the greater uncertainty in position?

   b Which has the greater uncertainty in momentum?

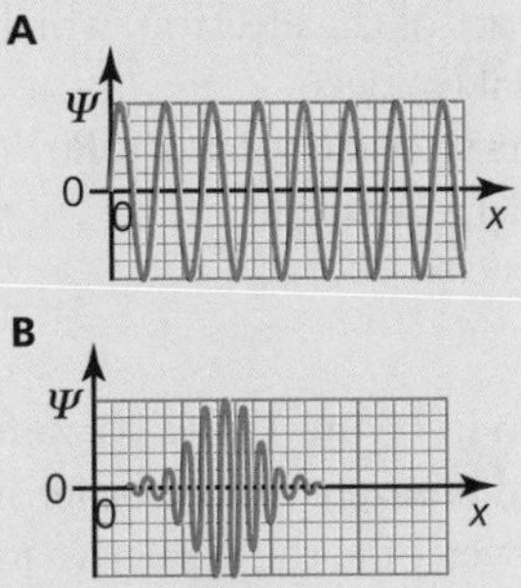

19 Consider example I above.

   a Show that the uncertainty in momentum is about 1×10^{-24} kg m s^{-1}.

   b Show that the kinetic energy of the electron is about 6×10^{-19} J.

   c Calculate the electric potential energy of the hydrogen atom.

   d Estimate (in eV) the overall energy of the atom in this state.

20 Consider example II above.

   a Determine the kinetic energy of an electron if it was constrained inside the nucleus of a hydrogen atom.

   b Explain, with the aid of a suitable calculation, why an electron with this kinetic energy cannot be restrained within such a small nucleus.

21 Confirm the uncertainty in energy quoted in example III above.

22 Consider again the pair production shown in Figure 12.11, for which we have said that the photon must have a minimum energy of about 1 MeV. An understanding of the uncertainty principle shows us that the pair could be produced from a much less energetic photon if they annihilated again a short time later. Calculate the maximum lifetime of a positron–electron pair produced in this way.

A

B

Figure 12.19

Common mistake

In a question, such as question 19, in which you are asked to confirm an approximate figure given in the question, you should quote your answer to more significant figures than those given (even if the physics does not justify them), before rounding to compare to the answer provided. Many students just give the same answer as was provided in the question.

Tunnelling, potential barrier and factors affecting tunnelling probability

Revised ☐

■ Potential barriers

■ Particles within atoms are constrained by attractive forces, for example negative electrons being attracted to positive nuclei. *Potential energy* is stored in these situations and we may describe the particles as being 'trapped' in potential wells, or behind potential barriers.

■ Figure 12.20 shows a two-dimensional representation of a (gravitational) potential well. A ball rolling on a friction-less surface like this may continually exchange kinetic energy and potential energy, but it cannot 'escape' unless it has sufficient kinetic energy.

■ The potential around a proton in a hydrogen atom can be similarly visualized. In classical physics an electron is trapped in the atom unless it has enough kinetic energy to get out of the potential well created by the electric force of

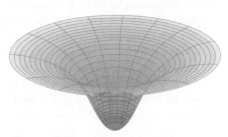

Figure 12.20

attraction. Figure 12.21 shows three simplified *one-dimensional* representations of an electron in a hydrogen atom trapped between rectangular (for simplicity) *potential barriers*.

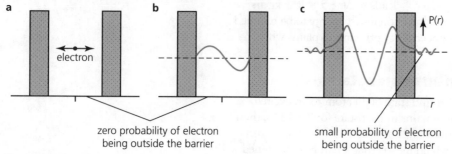

zero probability of electron being outside the barrier

small probability of electron being outside the barrier

Figure 12.21

- Figure 12.21a represents an electron as a *particle* moving around between potential barriers, but with insufficient kinetic energy to escape.
- Figure 12.21b recognizes the wave properties of the electron, which is drawn as a *standing wave* with nodes at the potential barriers.
- Figure 12.21c represents the electron by its *probability density*, $P(r)$. Note that this does not decrease to zero in the barrier, or outside.

Tunnelling

- The wave function for the electron shown in Figure 12.21c is *continuous through the potential barriers* around it, and non-zero on the other side. Classical physics tells us that the electron cannot overcome the attractive force, but in quantum physics there is a finite chance that an electron can *tunnel* through the barrier.
- The probability of **quantum tunnelling** depends on the mass and energy of the particle, and the 'height' and thickness of the barrier.
- Quantum tunnelling is only possible if the particle 'borrows' and 'repays' the extra energy required in a short enough time interval (Heisenberg's uncertainty principle).
- The *emission of alpha particles* from a nucleus is an interesting example of quantum tunnelling. Figure 12.22a represents the variation of potential energy around the nucleus from which an alpha particle was emitted.

> **Key concept**
>
> Quantum physics allows particles to do things that would be impossible under the principles of classical physics.

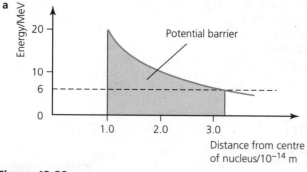

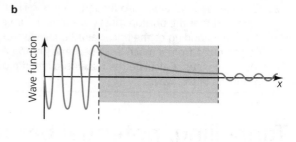

Figure 12.22

- In this example, an emitted alpha particle has kinetic energy of 6 MeV and is considered to be moving to the right. It is represented while in the nucleus on the left of Figure 12.22b by a free particle wave function with arbitrary units. By the principles of classical physics the alpha particle is trapped inside the nucleus because it does not have enough energy to overcome the potential barrier due to the electric attraction between the alpha particle and the rest of the nucleus: the energy required is shown by the red curve in Figure 12.22a. In this example it decreases from a maximum of 20 MeV, with increasing separation.
- If the same alpha particle with the same energy was 3.3×10^{-14} m from the nucleus, it would then be able to escape. The shaded area in Figure 12.22a represents the height and width of the potential barrier. Some alpha particles

will be emitted because the wave function is continuous through and after the barrier. The emitted free alpha particle wave function is shown on the right of Figure 12.22b.

■ The probability of alpha particle tunnelling can be linked to the half-life of the radioactive nuclide (see Section 12.2).

■ *Nuclear fusion* provides a contrasting example of a particle tunnelling *into* a barrier. Classical physics can be used to determine a theoretical minimum kinetic energy (and therefore temperature) necessary for protons to get close enough together for fusion to occur. Quantum physics explains why a small percentage of protons can do this at much lower temperatures.

QUESTIONS TO CHECK UNDERSTANDING

23 Explain the concept of quantum tunnelling.

24 If an alpha particle is emitted from a nucleus with kinetic energy which is 23 MeV less than the height of the potential barrier surrounding the nucleus, calculate the maximum time during which quantum tunnelling can occur.

25 Explain how it is possible that nuclear fusion can occur in the Sun, even when the temperature is lower than that predicted to be necessary by classical physics.

NATURE OF SCIENCE:

■ Observations; paradigm shifts

Observations and measurements of the line spectra emitted (or absorbed) by atoms are central to an understanding of atomic structure and the nature of electromagnetic radiation. Hydrogen is the simplest atom and it produces the simplest spectrum. An understanding of the hydrogen spectrum was the necessary first step before more complicated atomic spectra could be explained, and in the first 25 years of the twentieth century Bohr, and other scientists, devoted much time to observing and measuring spectra.

The introductions of wave–particle duality and quantum theory are two of many examples of paradigm shifts in physics and science: the accepted models and theories concerning an area of human knowledge are overturned, or at least significantly altered, as radical new ideas become widely accepted. Such shifts are commonly associated with major developments in scientific understanding, but it can be difficult for people to accept that previously held beliefs have to be seen as 'wrong'.

12.2 Nuclear physics

Revised ▢

Essential idea: The idea of discreteness that we met in the atomic world continues to exist in the nuclear world as well.

Rutherford scattering and nuclear radius

Revised ▢

■ The Rutherford–Geiger–Marsden experiment that led to the discovery of the nucleus was described in Section 7.3. In this section we will discuss the experiment in more mathematical detail.

■ In Section 7.3 we noted that the pattern of **Rutherford scattering** was explained by using Coulomb's law:

$$F_{E} = \frac{kq_{\alpha}q_{n}}{r^2}$$

where q_{α} is the charge on the alpha particle, q_{n} is the charge on the nucleus and r is the separation of their centres.

■ Figure 12.23 shows a two-dimensional gravitational analogue experiment which is often used to represent alpha particle scattering. Balls rolling on the hill move in a similar way to alpha particles approaching a nucleus. The *potential hill* is the opposite of a *potential well* (as discussed in Section 12.1), with its $\frac{1}{r}$ shape representing the potential (energy) changes when there is an inverse square law of repulsion between points.

Figure 12.23

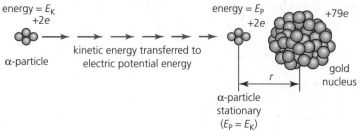

Figure 12.24

- Figure 12.24 shows an alpha particle (doubly positively charged) *directly* approaching a gold nucleus (with 79 positively charged protons). The alpha particle is decelerated by the repulsive force between positive charges and its kinetic energy is transferred to electric potential energy. For simplicity we will assume that the nucleus remains stationary.
- At some point the alpha particle's velocity will be momentarily reduced to zero, before it accelerated back the way it came. Because the alpha particle has considerable kinetic energy, we may assume that it gets very close to the nucleus: the closest separation of their centres is considered to be an *approximate* estimate for the radius of the nucleus.
- This separation can be determined using an energy consideration. Initial kinetic energy of alpha particle = electric potential energy, E_P, at closest point. For example, equating an alpha particle of energy 5.0 MeV to $E_P = \dfrac{kq_\alpha q_n}{r}$ (from Chapter 10), leads to an estimate of the radius of the gold nucleus (r) of about 5×10^{-14} m. This assumes that no significant amount of energy is transferred to the nucleus.
- But note that, although this is an important estimate, this simple calculation is based only on the use of the generic inverse square law of electric repulsion for *point* charges and the chosen energy of an alpha particle, rather than precise experimental data. For example, if less energetic alpha particles and/or nuclei with more charge were chosen for the calculation, a larger radius would be predicted (without experimental evidence).
- For more detailed information it is necessary to use alpha particles with greater energy, or bombarding particles which are not repelled (e.g. electrons – see below).

Explaining deviations from Rutherford scattering in high energy experiments

- If very high-energy alpha particles are used, they may actually enter nuclei and therefore be affected by the *nuclear strong forces*. Under such circumstances their interactions with nuclei cannot be explained simply by an inverse square law of electric repulsion.

> **Key concept**
> An alpha particle moving directly towards a nucleus will come to a momentary stop when all of its kinetic energy has been transferred to electric potential energy. The separation of their centres is then considered to be a estimate of nuclear radius.

> **Key concept**
> If alpha particles have enough energy to get very close to (or enter) the nucleus then their scattering will also be affected by the short-range strong nuclear force.

QUESTIONS TO CHECK UNDERSTANDING

26 A 7.2 MeV alpha particle is moving directly towards a stationary $^{204}_{82}$Pb nucleus.
   a What is the initial speed of the alpha particle (mass = 6.64×10^{-27} kg)?
   b Assuming that the nucleus cannot change position, what will be the smallest distance between their centres?
   c Explain why your answer may be considered as an estimate of the radius of the lead nucleus.
   d Explain why the alpha particle would not get so close to the nucleus if the lead atom was able to move freely.
   e If in an isolated system, the lead nucleus moved off with a speed of 1.8×10^6 m s^{-1}, determine the final velocity of the alpha particle.

27 a What is the approximate range of the strong nuclear force?
   b What types of particles are affected by this force?
   c In Rutherford scattering, why are only the very energetic alpha particles affected by the strong nuclear force?

Describing a scattering experiment including location of minimum intensity for the diffracted particles based on their de Broglie wavelength

- The **scattering of high-energy electrons** provides more accurate information about nuclear diameters than Rutherford scattering.
- Typically the electrons are accelerated by very high voltages, such that their wavelengths become comparable to nuclear diameters ($\approx 10^{-15}$ m) and diffraction effects are maximized.
- Electrons are much smaller than alpha particles and they are leptons, so that they are unaffected by the strong nuclear forces which affect high-energy alpha particles when they enter a nuclide, although they are affected by electric forces.
- We know that electrons behave as waves and the scattering process by nuclei has some similarities with the diffraction of light (photons) of wavelength λ by a narrow slit of width b, in which the first diffraction minimum occurs at an angle of $\sin \theta \approx \theta = \frac{\lambda}{b}$ (Section 9.2).
- Figure 12.25 shows a simplified version of the experimental arrangement. The intensity of the scattered electrons is measured for different angles. Figure 12.26 shows typical results.
- Measurement of the angle for the first diffraction minimum, θ, can be used to determine the nuclear diameter D: $\sin \theta \approx \frac{\lambda}{D}$, where λ is the de Broglie wavelength of the electrons used. The small angle approximation ($\sin \theta \approx \theta$) is usually *not* appropriate for use with electron scattering off nuclei (because the angles are usually too large).
- Experiments with different nuclei confirm that nuclear radii, R, (half the diameters) are proportional to the cube root of the number of nucleons, A. This conclusion is consistent with a nuclear model in which the nucleons are arranged close together, but remain separate, as shown in Figure 12.27. This model also suggests that nuclear density is the same for all nuclides (see below).
- Nuclear radii can be calculated from $R = R_0 A^{\frac{1}{3}}$, where R_0 is a constant known as the **Fermi radius** (= 1.2×10^{-15} m).

Nuclear density

- The **nuclear density** of a particular nuclide can be determined from density $= \frac{\text{mass}}{\text{volume}}$. Experimental results suggest that nuclei are spherical in shape, so that their volume, $V = \left(\frac{4}{3}\right)\pi R^3$.

- All nuclear densities are approximately equal (2×10^{17} kg m^{-3}). The density of nuclei is equivalent to a cubic centimetre having a mass about equal to that of a large mountain.
- The only macroscopic objects with a density as large as nuclei are **neutron stars** and *black holes* (collapsed massive stars).

Electrons with such high energies will have masses significantly greater than their rest masses. This is a relativistic effect, as discussed in Option A.

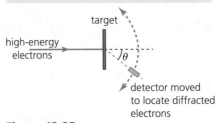

Figure 12.25

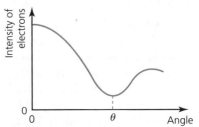

Figure 12.26

Key concept

High energy electrons have very short wavelengths and are diffracted by nuclei. The angle of the first diffraction minimum can be used to determine nuclear diameter.

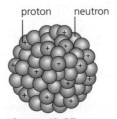

Figure 12.27

QUESTIONS TO CHECK UNDERSTANDING

28 a What is the momentum of an electron which has a de Broglie wavelength of 1.0×10^{-15} m?

   b Calculate the kinetic energy in MeV of this electron.

   c What p.d. is needed to accelerate the electron from rest to this energy?

29 When electrons of wavelength 2.7×10^{-15} m were incident on iron nuclei, the first scattering minimum was detected at an angle of 20°.

   a What is the approximate diameter of an iron nucleus?

   b If the p.d. accelerating the electrons was increased, what would happen to the scattering angle?

30 a Explain why the experimental fact that nuclear radii are proportional to the cube root of their nucleon number suggests that the nucleons are separate, but closely packed together.

b Suggest why it is reasonable to suggest that nuclei are approximately spherical in shape.

31 a What is the radius of a $^{63}_{29}$Cu nucleus?

b Confirm that the density of this nucleus $\approx 2 \times 10^{17}$ kg m^{-3}.

32 An atomic nucleus has a radius of 3.6×10^{-15} m. Suggest what element this might be.

33 As their name suggests, neutron stars are composed mostly of neutrons. If a neutron star has a radius of 20 km, estimate:

a the number of neutrons it contains

b the total mass of the star.

(Assume that the radius of a neutron is 1.2×10^{-15} m.)

Nuclear energy levels

Revised ☐

- In Chapter 7 we saw that there are *discrete energy levels* within atoms. The concept of the quantization of energy on the microscopic scale also applies to nuclei.

Describing experimental evidence for nuclear energy levels

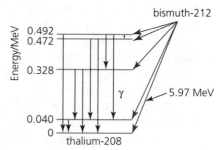

Figure 12.28

- The *alpha particles* and *gamma rays* that are emitted from any particular unstable nuclei always have exactly the same discrete energies.
- The blue arrows on the right in Figure 12.28 show five possible alpha particle decays of bismuth-212. The alpha particle with the most energy (5.97 MeV) is emitted when the daughter nucleus (thallium-208) is left in its ground state. There are four other possible energies for alpha particles emitted from bismuth-212, each of which leaves the thallium nucleus in an excited state.
- When the excited thallium nucleus decays to its ground state, one or more gamma rays photons will be emitted. The photon energies emitted due to **nuclear transitions** are much greater than the photon energies released by electron transitions.
- When an alpha particle is emitted, the nucleus will recoil in exactly the opposite direction (conservation of momentum in one direction). See Figure 12.29.

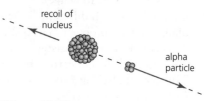

Figure 12.29

- But the emission of beta particles from unstable nuclei (Section 7.1) is more complicated because three particles are involved, see Figure 12.30. The three particles may move in different directions and applying the law of conservation of momentum in three dimensions shows that continuous ranges of all particle velocities are possible. This explains why **beta particle spectra** are observed.

Three particles can move apart in any combination of directions and share the available energy in a continuous range of different ways.

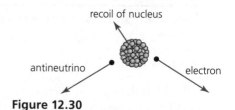

Figure 12.30

> **Key concept**
>
> The discrete nature of emitted alpha and gamma energies is evidence for the existence of discrete **nuclear energy levels** because the emissions always arise as consequences of changes between the same levels.

The neutrino

Revised ☐

- From Section 7.1 we know that neutrinos or antineutrinos are emitted from a nucleus in the process of beta particle decay, see Figure 12.30:

beta-negative decay   beta-positive decay

$n \rightarrow p + e^- + \bar{v}_e$ $p \rightarrow n + e^+ + v_e$

- The existence of neutrinos and antineutrinos was first proposed to explain why emitted beta particles have continuous ranges of energies (see above).

> **Key concept**
>
> Neutrinos have no charge, *very* small mass and travel at speeds close to the speed of light. Although they are believed to be among the most common particles in the universe, they are very penetrating and difficult to detect because interactions with other particles are very, very rare.

The law of radioactive decay and the decay constant

Revised ☐

- In Section 7.1 we discussed how the count rate from a radioactive source varies with time. Graphical examples were given and the rate of decay characterized by the concept of *half-life*, $T_{1/2}$. See Figure 7.12 (Chapter 7). A dice analogue experiment was used to help explain the shape of the graphs.

- A count-rate from a pure radioactive source decreases *exponentially* because the rate of decay of nuclide depends on the number left undecayed (which itself reduces with time).

- As discussed in section 11.3, an exponential decrease (decay) occurs when the rate of change of any quantity N, $-\frac{\Delta N}{\Delta t}$, is proportional to the number remaining at that moment. $\frac{\Delta N}{\Delta t} = -\lambda N$, where λ is a constant. For radioactivity, this is known as the **law of radioactive decay** and λ is known as the **decay constant** (unit: s^{-1}).

- The decay constant is defined as the probability of decay per unit time. The larger the value of the decay constant, the more rapid the radioactive decay and the shorter the half-life.

- In radioactivity, $\frac{\Delta N}{\Delta t}$ is known as the **activity** (A) of the source. $A = \lambda N$. One decay per second is called an activity of 1 **bequerel**, Bq.

- A received count rate is *not* equal to the activity of the source, but may often be considered to be proportional to it, so that count rates can often be used in calculations instead of activities.

> **Key concepts**
>
> The *activity*, A, of a radioactive source is the number of decays per second ($= \frac{\Delta N}{\Delta t}$). Activity is proportional to the number of undecayed atoms at that time, N. The rate of decay is represented by the *decay constant*, λ.
>
> $A = \frac{\Delta N}{\Delta t} = -\lambda N$.
>
> $\lambda = \ln \frac{2}{T^{1/2}}$

■ Exponential equations

- In Section 11.3 we saw that $N = N_0 e^{-kt}$ is a mathematical equation that describes *any* exponential decrease: a quantity falls from a value N_0 to N in a time t and the rate of decrease is controlled by the size of the constant k (for radioactive decay λ is used instead of k, as described above).

- Taking logarithms of $N = N_0 e^{-\lambda t}$ to the base e (ln) gives $\ln N = \ln N_0 - \lambda t$.

- The time taken to fall from N_0 to $\frac{N_0}{2}$ is the half-life, $T_{1/2}$, and inserting these values in the highlighted equation leads to $T_{1/2} = \frac{0.693}{\lambda}$ (Since $\ln 2 = 0.693$).

- Since the activity, A, of a source is directly proportional to N, it follows that $A = A_0 e^{-\lambda t}$, where A_0 is the initial activity of the source. Variations in count rates may also be calculated in the same way.

- Because $A = \lambda N$, the activity at a time t after the number of undecayed atoms was N_0 can be determined from $A = \lambda N_0 e^{-\lambda t}$.

> **Key concept**
>
> The number N of undecayed nuclei in a radioactive sample after time t is given by the equation $N = N_0 e^{-\lambda t}$, where N_0 is the number of undecayed nuclei at the start of timing.

■ Solving problems involving the radioactive decay law for arbitrary time intervals

QUESTIONS TO CHECK UNDERSTANDING

37 The isotope cobalt-60 is often used in schools for gamma ray demonstrations. It has a half-life of 5.27 years. After 2 years in a school the activity of the source has decreased to 1.5×10^5 Bq.

   a Calculate the number of undecayed cobalt-60 atoms in the source.

   b Determine the activity of the source when it was first delivered to the school.

38 Technetium-99 is used widely for medical diagnoses. It has a half-life of 6.0 h.

   a What is the decay constant of this isotope?

   b If some of this isotope, with an activity of 800 MBq was injected into a patient (see Figure 12.31) at 14.00 in the afternoon, what would its activity be at 18.00 on the next day?

Figure 12.31

39 The number of unstable nuclei in a rock sample of a uranium isotope is believed to have decreased from 2.50×10^{17} to 2.25×10^{17} in a time of 680 million years. Estimate the half-life of this isotope.

40 An isotope of protactinium was used in a school laboratory for half-life determination. The count rate fell from $72\,s^{-1}$ to $7\,s^{-1}$ in 5 minutes. If the average background count was $1\,s^{-1}$, estimate the half-life of the isotope.

■ Explaining the methods for measuring short and long half-lives

■ As discussed in Section 7.1, the half-life of some isotopes can be found by measuring count rates over a suitable period of time. (The count rates should be adjusted for the background count.) For this method to be possible, the count rate must decrease significantly in the time available. For example, if the experiment has to be completed in one hour, the half-life being determined should probably be no more than two hours.

Expert tips

Radioactivity calculations in this course involve pure sources. In practice, nuclides may decay into other unstable nuclides (producing a decay 'chain'), so that sources often contain mixtures of radioactive nuclides, although the activity may be dominated by the decay of one particular nuclide.

Figure 12.32 shows one way of measuring the radiation emitted by just atoms of just one particular nuclide (protactinium). The sealed strong plastic bottle contains two layers of different liquids. The lower layer contains a uranium isotope and all of its decay products (including protactinium). The upper layer contains an organic solvent. When the liquids are mixed by shaking, any protactinium present is removed by the organic solvent. The liquids are allowed to separate again before measurements are started.

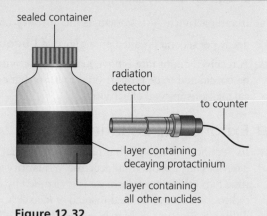

Figure 12.32

■ In Section 7.1 we showed how a value for a half-life could be determined from a graph of count rate–time. We are now also able to use the exponential equation $A = A_0 e^{-\lambda t}$ with two count rates (instead of activities, A) and the time, t, between them to determine a value for the decay constant, λ. The half-life can then be calculated from $T_{1/2} = \dfrac{0.693}{\lambda}$.

- Drawing a graph will improve the accuracy of the half-life determination: since $\ln A = \ln A_0 - \lambda t$, a graph of ln A–t (using count rates), will have a gradient of $-\lambda$ (compare with $y = mx + c$), from which the half-life can be determined. See Figure 12.33.

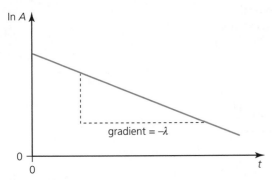

Figure 12.33

- *If a radionuclide has a long half-life*, then its activity will be effectively constant over a measurable period of time, so that the previously described method cannot be used.
- Instead, the number of nuclei, N, in a *pure* sample can be calculated from its mass, its molar mass and Avogadro's constant $\left(N = \frac{mN_A}{\text{molar mass}}\right)$. The measured activity, $A = -\lambda N$ can then be used to calculate the decay constant.
  - It is probable that a sample of an isotope with a long half-life will have a very low activity, and this may reduce the accuracy of the experiment, especially if the magnitude of the count is comparable to the magnitude of the background count.
  - This method requires measurement of the actual *activity* of the source, rather than just a count rate. Figure 12.34 explains the difference.
  - The ratio $\frac{4\pi r^2}{\text{area of detector}}$ can be used to determine the total count that would be due to radiation passing through a spherical surface of radius r. This can be assumed to be equal to the activity if no radiation is absorbed between the source, and the detector records all the radiation that enters it.

Expert tip

The random variation of a radioactivity count N is not experimental error or uncertainty. It is purely a consequence of the random nature of the decay process, and it is most noticeable with small counts. A count of N is often assumed to vary by approximately $\pm\sqrt{N}$. This means that the percentage uncertainty is much greater for small counts. For example, a small count rate of 25 min^{-1} could be expected to vary between 20 min^{-1} and 30 min^{-1} (±20%). Whereas a count rate of 2500 min^{-1} ±50 min^{-1} has only a percentage variation of 2%. Low count rates are difficult to distinguish from similar background rates.

Key concept

Determining a value for a very long half-life involves measuring very low activities and the mass of the isotope involved.

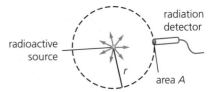

Figure 12.34

QUESTIONS TO CHECK UNDERSTANDING

41 The count rate detected from a radioactive nuclide decreased from 48 s^{-1} to 34 s^{-1} in 1.50 minutes.

a Determine the half-life of this nuclide.

b Sketch a graph to show how the count rate changes over a time of 6.0 minutes (starting at 48 s^{-1}).

c Draw a ln (count rate)–time graph for the same time.

42 Figure 12.35 shows a ln$A$–t graph for the decay of a pure isotope.

a What was the activity of the source at time $t = 0$?

b What is the half-life of the isotope?

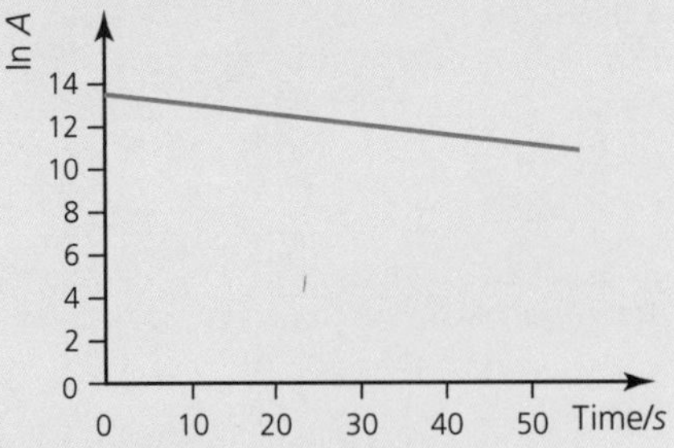

Figure 12.35

43 A nuclide has a half-life of 2.0 hours. By what percentage does its activity decrease in one hour?

44 A radiation detector with a receiving area of 1.2 cm^2 is placed 8.4 cm from a source of activity 6.3×10^4 Bq.

a Calculate the maximum count rate that could be detected.

b Explain why your answer is a *maximum* value.

c Compare your answer to a typical background count of 0.5 Bq.

45 a How many atoms are present in 1.0 μg of carbon-14?

b Carbon-14 has a half-life of 5730 years. What is the decay constant for this nuclide?

c Determine the activity from 1.0 μg of pure carbon-14.

46 Calculate the activity from a 5.0 mg source of pure radium-226. Its half-life is 1600 years.

■ Theoretical advances and inspiration

Discoveries in atomic and nuclear physics have often only occurred because physicists were inspired by unexplained observations to make new theoretical predictions. The proposed existence and properties of neutrinos is a good example of this method of scientific advance. Experiments were then devised specifically to test those predictions.

■ Advances in instrumentation

Larger and more powerful particle accelerators together with improved particle detection technology have been a major driving force behind recent advances in nuclear physics.

■ Modern computing power

Enormous amounts of data are collected automatically from particle and radiation detectors of various kinds in nuclear physics experiments around the world. The analysis of all this information requires computing power that would not have been possible until recent years.